You CAN Teach Yourself
Basic Algebra

Roderick J. MacGregor

Southern Maine Community College

ISBN-13: 978-1505992861

ISBN-10: 1505992869

Introduction

This book is not meant to be a textbook, nor is it a replacement. It is less complete and more informal. Its purpose is to help clarify the logical extension of algebra as part of the mathematical system that we already understand. Although we will discuss various algebraic manipulations that can be used for solving problems, the main objective will be to help you understand and be able to use this potent extension of arithmetic.

Algebra follows the same structure as the arithmetic you have utilized throughout your life. It extends the structure to include the concepts of variables and equations and their manipulation to make arithmetic so much more versatile and useful in our increasingly complex world.

Throughout this book, we will confirm that the various procedures we use with algebraic expressions and equations are mathematically correct. This means that they follow the same rules as arithmetic with which you are already familiar. As the concept of variable is more abstract, we will often employ parallel examples with the variables replaced by numbers to clarify that the algebraic manipulations that we use are in fact following established arithmetic rules. This will help you understand *why* a particular manipulation is mathematically sound, and thereby promote a deeper understanding that will allow you to have confidence in the correctness of your solutions to the various kinds of problems you encounter.

In order to be proficient in the use of algebra, we must first understand the structure of arithmetic and its operations. You are familiar with these operations since we use them every day: *addition, subtraction, multiplication, division*. We also will review the operations of raising numbers to powers (exponentials) and its associated operation, finding roots of numbers (radicals).

If you diligently work on the **Concept Homework** as well as the **Exercises** at the back of each section, you will deepen your understanding of Algebra and its usefulness.

Table of Contents

4

Topic 1 – Rules for Signed Number Operations

In the beginning years of elementary school, we all learned the rules for addition, subtraction, multiplication and division of **whole numbers** [0, 1, 2, 3, ...]. Negative numbers were not addressed. Later on, they were added to the set of whole numbers to make up the set of **integers** $\left[.....-4,-3,-2,-1,0,1,2,3,4.....\right]$. In this discussion we will review how "signed" numbers **[positive or negative]** impact these basic **operations** of addition, subtraction, multiplication and division.

Every number has a sign, *even if it is not expressly written!* For example, the number 4 is understood to be **positive** 4. If we wish to express a negative number we will often put it in parenthesis. For example $\left(-7\right)$, means that the 7 is negative.

> **Examples:** (a) $3+4$ means "positive 3" *add* "positive 4".
>
> (b) $3-4$ means "positive 3" *subtract* "**positive** 4".
>
> (c) $3+\left(-4\right)$ means "positive 3" *add* "negative 4"
>
> (d) $3-\left(-4\right)$ means "positive 3 *subtract* "negative 4"
>
> (e) $-3-4$ means "negative 3" *subtract* "**positive 4**".

The parenthesis in example (c) is used to indicate the negative symbol "−" is attached to the number 4 and is not the **operational symbol** of subtraction. In example (d) $3-\left(-4\right)$, the first "−" symbol indicates the **operation** of subtraction and the "−" symbol within the parenthesis indicates the number 4 is negative. In mathematics we use the same symbol, "−", to indicate *both* the **operation** of subtraction as well as the **concept of a negative number**. The parenthesis is used to distinguish between the two usages.

To further complicate things, in example (e) "−3" means "negative 3" even though it is not in parenthesis. The parenthesis is left out there because there is no number preceding it. It could not be a subtraction operation as there is no number from which 3 is being subtracted.

THE ADDITION OPERATION

Since we have to deal with negative as well as positive numbers **[integers]**, we must have rules for addition that will be **consistent** for both positive and negative numbers.

At this point we need to discuss the concept of **absolute value.** As you know, all real numbers, except 0, can be either **positive** or **negative.** When we talk about the **absolute value** of a number, we look only at its **magnitude** [size] without considering it's sign. *All absolute values are considered to be positive*. In mathematics, absolute values are indicated by putting the number in between two **elongated vertical** [straight up and down] **lines**. **Examples:** $|3|$ or $|-7|$

Since all absolute values are positive, **we can remove** a number from an absolute value symbol, $|\ |$, by making the value inside it positive.

Examples: $|3| = 3$ or $|-7| = 7$

Another way of expressing this concept is thinking of absolute value as the **distance from the number to** 0 **on the number line**. In thinking about distances, we realize that **they must be positive**. For example, saying that the length of a pencil [a distance measure] is -7 inches **makes no sense.** A distance can't be less than 0. In the context of the number line, 3 is 3 units of distance from 0 on a number line measuring from the right side of 0. Therefore its absolute value is 3. Numbers on the left side of 0 are negative. -7, for example, is 7 units from 0, measuring from the left side. Therefore its absolute value is 7. We need the concept of absolute value to fully understand the **rules for addition of signed numbers**:

Rules for the Addition Operation

1. If the **sign** [positive or negative] of the numbers being added **are the same**:
   A. Add their **absolute values**
   B. Then keep the **sign** that both numbers have.

    Example 1. $3 + 5 = 8$

Although this seems obvious since we have been adding positive numbers since childhood without thinking about their being positive, we now need to think of this operation in the context of the mathematical rule for **signed numbers**:

(A) Both numbers are positive. Therefore we add their absolute their values which in this case is 8.
(B) Keep the **sign** of the numbers being added, which is positive.

    Therefore, $3 + 5 = 8$

Example 2. $\left(-3\right)+\left(-6\right) = -9$

(A) Adding the **absolute values** of these numbers results in 9. $[3+6=9]$

(B) Keep the **sign** of the numbers being added, which is negative.

Therefore, $\left(-3\right)+\left(-6\right) = -9$.

2. If the **signs** of the numbers being added are **opposites** [one positive and one negative]:

A. Subtract the smaller absolute value from the larger absolute value.

B. Keep the **sign** of the number with the larger absolute value.

Example 1. $-6+7$

6 is a the smaller absolute value. Therefore, we subtract it from 7 which is a larger absolute value, which results is 1. The result is **positive** since we keep the **sign** of the number that has the larger absolute value, which in this case is 7 .

Therefore, $-6+7 = 1$

Example 2. $9+\left(-10\right)$

9 is the smaller absolute value. Therefore, we subtract it from 10 which is the larger absolute value which results in 1. The result is **negative** since we keep the **sign** of the number that has the larger absolute value, -10 .

Therefore, $9+\left(-10\right) = -1$

This might seem "a long way home" for many of you, who *instinctively* can arrive at the correct answer. However, it is important to understand the rules **[or logical basis]** for what we are doing, as this is the key for understanding and using algebra correctly. This idea parallels *instinctively* knowing what is proper grammatical usage in a language. We should **also know the rules of grammar** for that language.

THE SUBTRACTION OPERATION.

The only rule for subtraction is: ***CHANGE THE PROBLEM TO ADDITION!***

How?
   A. Change the subtraction **operation symbol** to an addition **operation symbol**.
   B. Reverse the **sign** of the number **following** the subtraction operation symbol.

Example 1: $5-3$

Note that the connecting symbol between 5 and 3 is the **subtraction operation**. 3 is considered a **positive number!**

(A) Using the above rule: $5-3 \Rightarrow 5+(-3)$

We have **changed the subtraction operation to addition** by changing 3 [The number following the subtraction symbol] to -3.

(B) Now **we use the rules for addition with opposite signs** for $5+(-3)$.

Subtract the smaller absolute value from the larger absolute value and keep the sign of the larger absolute value: $5+(-3)=2$

Therefore: $5-3=2$ [solution]

Again, this might seem "a long way home" since we learned very early on that $5-3=2$. However, it is important to see this example in the context of signed numbers. By doing this, we can see that what we first learned about subtraction of only positive numbers is consistent with the rules for both positive **and** negative numbers [integers].

Example 2: $3-5$

(A) We change the subtraction operation to addition: $3-5 \Rightarrow 3+(-5)$

(B) Now **we use the rules for addition with opposite signs for** $3+(-5)$.

Subtract the smaller absolute value from the larger absolute value and keep the sign of the larger absolute value: $3+(-5)=-2$

Therefore: $3-5=-2$ [solution]

Example 3: $\qquad$ $-3-5$

(A) We change the subtraction operation to addition: $\qquad$ $-3-5 \Rightarrow -3+(-5)$

[The negative symbol in front of the 3 does not indicate subtraction since there is nothing preceding it to subtract it from!]

(B) **Now we use the rules for addition with same signs:**

Add the absolute values and keep the sign of the numbers being added:

$$-3+(-5) = -8$$

Therefore, the solution to the original example: $\qquad$ $-3-5 = -8$ [solution]

Example 4: $\quad$ $-3-(-5)$

(A) We change the subtraction operation to addition: $-3+5$

(B) Now **we use the rules for addition with opposite signs**:

Subtract the smaller absolute value from the larger absolute value and keep the sign of the larger absolute value: $\qquad$ $-3+5 = 2$

Therefore: $\qquad$ $-3-(-5) = 2$ [solution]

At this point we should **generalize** this concept of changing addition to subtraction using algebraic variables:

$$a - b = a + (-b)$$

where a and b can represent any real numbers.

THE MULTIPLICATION/DIVISION OPERATIONS

Just as addition and subtraction are related [you change subtraction to addition by changing the sign of the number being subtracted], the multiplication and division operations are also related. The concept of multiplication came about to express repeated addition of the same number a certain amount of times. For example, if we are adding $5+5+5+5+5+5$, we could go through the series of additions arriving at 30. **Since the number being added is the same** we can conceptualize this calculation as a multiplication operation: 5 taken 6 times results in 30. Translating this from words to mathematics, the calculation is written $5 \cdot 6 = 30$.

Division is the reverse of this operation. In this example, we would start with the result of the multiplication, 30. Then we would ask what number would divide 30 into 6 equal parts? The answer would be the number 5. Mathematically, we would express this operation as $30 \div 6 = 5$.

So, now we can see the relationship between the multiplication operation and the division operation. Using this example, we can say that if $5 \cdot 6 = 30$ then $30 \div 6 = 5$. If fact, this relationship will *always* hold, regardless of what numbers are being multiplied. To express this generalized relationship, we turn to **algebraic variables**:

If $a \cdot b = c$, **then** $c \div b = a$, where a and b can represent *any* real numbers!

In algebra, it is not necessary to use a dot between variables to indicate multiplication. We can leave out the dot and write this relationship as:

If $ab = c$, **then** $c \div b = a$

[The *implied* operation between ab is multiplication.]

While we are on the subject of notation, it would be a good time mention that division can also be expressed as a **fraction**: $c \div a = b$ can be expressed as $\dfrac{c}{a} = b$

where $\dfrac{c \text{ is the dividend - the number being divided}}{a \text{ is the divisor - the number doing the dividing}}$

So, algebraically, this relationship between multiplication and division can be expressed as:

If $ab = c$, **then** $\dfrac{c}{b} = a$

To make this generalization more concrete, supply the numbers of your choice for a and b. Then use them to show that the above generalization is correct.

Rules for Multiplication and Division of Signed Numbers

1. If the **signs** of the numbers being **multiplied** or **divided** are the same [either both positive or both negative]: The result is **always positive**.

Example 1: $(-3)(-5)$

[Parentheses, ( ) ( ), without an operation symbol between them indicate multiplication.]

The **signs** of the numbers being multiplied are the same, so the result is positive.

Therefore, $(-3)(-5) = 15$

Example 2: $2 \cdot 3$ [A dot, $\cdot$, also indicates multiplication.]

The **signs** of the numbers being multiplied are the same, so the result is positive.

Therefore, $2 \cdot 3 = 6$

Example 3: $-8 \div (-2)$. This division can be represented as a fraction:

$$\frac{-8 \quad \text{[dividend]}}{-2 \quad \text{[divisor]}}$$

[The dividend is the number being divided (numerator). The *divisor* is the number doing the dividing (denominator).]

The **signs** of the numbers being divided are the same, so the result is positive.

Therefore, $-8 \div (-2) \Rightarrow \dfrac{-8}{-2} \Rightarrow 4$

Example 4: $10 \div 5$

The **signs** of the numbers being divided are the same, so the result is positive.

Therefore, $10 \div 5 \Rightarrow \dfrac{10}{5} \Rightarrow 2$

Example 5: $\dfrac{-10}{-5}$ [A fraction indicates a division operation.]

The *signs* of the numbers being divided are the **same**, so the result is positive.

Therefore, $\dfrac{-10}{-5} = 2$

2. If the signs of the numbers being multiplied or divided are opposite
[one positive and one negative]: the result will **always be negative.**

Example 1: $-3 \cdot 5$ or $(-3)(5)$

The **signs** of the numbers being multiplied are **opposite. Therefore the result is negative.**

Therefore, $-3 \cdot 5 = -15$ or $(-3)(5) = -15$

Example 2: $2(-3)$

The **signs** of the numbers being multiplied are **opposite. Therefore the result is negative.**

[Since there is no addition or subtraction symbol between 2 and (-3) the multiplication operation is implied.]

Therefore, $2(-3) = -6$

Example 3: $8 \div (-2)$ or $\dfrac{8}{-2}$

The **signs** of the numbers being divided are **opposite. Therefore the result is negative.**

Therefore, $8 \div (-2) = -4$ or $\dfrac{8}{-2} = -4$

Example 4: $-8 \div 2$ or $\dfrac{-8}{2}$

The **signs** of the numbers being divided are **opposite. Therefore the result is negative.**

Therefore, $-8 \div 2 = -4$ or $\dfrac{-8}{2} = -4$

Here is an interesting question whose answer is utilized often in algebra:

Are the following fractions *equivalent* [have the same quantitative value]?

$$\frac{8}{-2} \quad \text{and} \quad \frac{-8}{2}$$

As can be seen from examples 3 and 4, both fractions are equal to -4. Therefore, they are *equivalent* and therefore ***interchangeable***. This is true for **any such fractions.**

$$\text{For example: } \frac{1}{-2} = \frac{-1}{2} \;;\quad \frac{-3}{4} = \frac{3}{-4} \;;\quad \frac{7}{-1} = \frac{-7}{1}.$$

If we generalize this concept where a and b can take on any real values, it is true that:

$$\frac{-a}{b} = \frac{a}{-b}$$

If a fraction represents a negative value, the negative sign is often put in front of the fraction.

Since $\frac{-a}{b}$ represents a negative value **[The *signs* of the numbers being divided are *opposite*. Therefore the result is always negative!]** and $\frac{a}{-b}$ represents a negative value **[for the same reason]**, we can extend the concept above:

$$\frac{-a}{b} = \frac{a}{-b} = -\frac{a}{b}$$

Here is a song (sung to row, row, row your boat) that summarizes these rules:

Same signs add and keep,
Different signs subtract.
Keep the sign of the higher number **[absolute value]**,
Then it will be exact.

14

Concept Homework

As an assessment of your understanding of the concepts set forth in this section, answer the following questions. If necessary, review the material to help you arrive at the correct conclusions. With "true or false" statements, if the statement is true, make up an example that supports the statement without using examples given in the material. If a statement is false, correct the wording to make it true. Then make up an example that supports the statement without using examples given in the material.

True or False:

a. Every number has a **sign** which is distinct from an addition and subtraction **operation symbol.**

b. When **adding** numbers with **opposite signs**, the result is *always negative*.

c. A **subtraction operation** can always be changed to an *addition operation* by **reversing the sign** of the number following the subtraction symbol [the number being subtracted].

d. The division operation can be expressed in the form of a fraction.

e. Division and multiplication of opposite signed numbers always results in a **negative** number.

f. When **adding** numbers with the **same sign** [either both positive or both negative], the result will also have **that sign**.

g. When **multiplying or dividing** numbers with **same sign** [either both positive or both negative], the results will also have **that sign**.

h. If two fractions are **interchangeable**, they must have the same quantitative value.

i. $\dfrac{-2}{3}$ has the same quantitative value as $\dfrac{2}{-3}$ and both values are negative.

j. $-\dfrac{-2}{-3}$ has the same quantitative value as $\dfrac{2}{-3}$.

Solutions: a. True; b. False; c. True; d. True; e. True; f. True; g. False; h. True; i. True; j. True

Rationale for j. being true: $-\dfrac{-2}{-3} \Rightarrow -\dfrac{2}{3} \Rightarrow \dfrac{2}{-3}$

Exercises:

1. Perform the indicated operations:

a. $7 - 9 =$ b. $-55 + (-8) =$ c. $18 - (-5) =$ d. $-9 + (-2) =$

e. $-201 - (-23) =$ f. $0 - 32 =$ g. $0 - (-5) =$ h. $8 - (-2) - 10 - 3 =$

2. Perform the indicated operations:

a. $-9 \cdot 8 =$ b. $(-6)(-5) =$ c. $-36 \div 6 =$ d. $\dfrac{-42}{-7} =$ e. $(-7)(-7) =$

f. $(6)(-5)(5) =$ g. $-8 \div (-2) =$ h. $(0)(-7)(-9) =$ i. $\dfrac{0}{-10} =$

Solutions: 1a. −2; 1b. −63; 1c. 23; 1d. −11; ; 1e. −178; 1f. −32 ; 1g. 5; 1h. −3; 2a. −72; 2b. 30; 2c. −6; 2d. 6; 2e. 49;

2f. −150; 2g. 4; 2h. 0; 2i. 0.

Topic 2 - The Order of Operations

Many times when we are simplifying arithmetic or algebraic expressions, there are a series of operations to be performed [a series of numbers connected by more than one operation]. There has to be a way of deciding which operation to perform first, second, third, etc. The agreed upon guide that is universally accepted is known as **The Order of Operations**.

Most of you have heard of the acronym [each letter represents the first letter of a word] PEMDAS. Sometimes students were taught to remember this acronym with:

"**P**lease **E**xcuse **M**y **D**ear **A**unt **S**ally"

PEMDAS represents the order in which we should do mathematical operations when there is a multi-operational expression.   **Example:** $2-[3-(2-3)^2 \cdot 2]+5$

As can be seen in this very complicated example, there are many **operations** going on here: **Exponents, Multiplication, Addition and Subtraction**

Also in this example, there are operations **within a bracket** and a **parenthesis within that bracket.** The question becomes "In what order should we do these various operations?" This is important because changing the order in which we do the various operations could lead to different results using the exact same numbers! If everyone performing this series of operations didn't get the same solution, then our **mathematical system would be seriously flawed**.

Therefore, the world has agreed on a order in which to approach such operations so that *there will only be one possible solution when such expressions are executed.* This system in **boiled down form** is called **PEMDAS**. This acronym stands for the order in which we should perform these various operations:

 First: **P**arenthesis - actually refers to all **grouping symbols**: brackets [ ], parentheses ( ), and absolute value symbols | |. Perform all operations that are within these grouping symbols first. **Work from the inside out**.

Next: **E**xponents   [numbers raised to a power]

Next: **M**ultiplication **OR** Division [as it appears from left to right in the expression]

 Note: It is important to know that multiplication is **NOT always** done before division. multiplication and division are done **in the order in which they appear in the problem in left to right order.**

Last: **A**ddition or **S**ubtraction   [as it appears from left to right in the expression]

 Note: It is important to know that addition is **NOT** always done before subtraction. Addition and subtraction are done **in the order in which they appear in the problem in left to right order.**

The easiest way to correctly perform problems involving order of operations is to:

    (1) Work vertically [downwards] with each step and

    (2) Do one step at a time.

Let's go back to the example given above and see how this problem should be attacked.

Example: $\qquad\qquad\qquad\qquad\qquad\qquad\qquad\qquad$ $2-[3-(2-3)^2 \cdot 2]+5$

As operations within **Parentheses [and also brackets]** are to be **executed first** from **inside out**, We start with the **inside grouping symbol**: the

parenthesis within the brackets : $\quad 2-[3-\underline{(2-3)}^2 \cdot 2]+5$

Notice that I resist the temptation to do more than one operation on each step!

We first perform the subtraction operation that is **within the parenthesis**

And inside the brackets: $\quad 2-[3-\underline{(-1)}^2 \cdot 2]+5$

Now we concern ourselves with operations within the **brackets** . The **exponential operation** is the highest order and is performed next.

$(-1)^2 \Rightarrow -1 \cdot -1 = 1$, so the expression now becomes: $\quad 2-[3-\underline{1 \cdot 2}]+5$

Still within the brackets, we must perform the multiplication operation next: $\quad 2-[\underline{3-2}]+5$

Still within the parenthesis, we perform the subtraction: $\quad 2-1+5$

There is now longer a need for the brackets, since all the operations within them have been performed.

We are now down to just addition and subtraction operations. **These should always be the last operations (once all grouping symbols have been eliminated) that we perform**. Since addition and subtraction are performed as they occur in the expression from left to right, we will do the

subtraction operation first: $\qquad \underline{2-1}+5$

which results in: $\qquad 1+5$

And finally, we perform the last addition operation: $\qquad 6$ [solution]

By following these rules carefully [and doing the operations correctly], *anyone* **in any country in the world, or any computer in the world** _must_ **come to the same solution!**

You are probably thinking that this is too long and too much work. However, this **organized** and **careful** approach to this complex problem leads to the correct solution. If you try to combine steps to save time, you risk making a fatal error that leads to an incorrect answer.

Concept Homework

As an assessment of your understanding of the concepts set forth in this section, answer the following questions. If necessary, review the material to help you arrive at the correct conclusions. With "true or false" statements, if the statement is true, make up an example that supports the statement without using examples given in the material. If a statement is false, correct the wording to make it true. Then make up an example that supports the statement without using examples given in the material.

True of False

1. Order of operations dictates that multiplication is always done before division in a simplification involving multiple operations.

2. The "P" in PEMDAS stands for parenthesis but actually includes all kinds of grouping symbols **[parentheses, absolute value symbols, and brackets]** .

3. If there are grouping symbols within other grouping symbols, the proper procedure is to perform the operation(s) that are inside the outside grouping symbols first. **[Work from the inside out.]**

4. We never perform addition or subtraction before multiplication, even if the operation is within a parenthesis.

5. It is a good strategy to perform as many operations as we can simultaneously when simplifying complex order of operations problems.

Solutions: 1. False; 2. True; 3. True; 4. False; 5. False.

Exercises

Simplify:

1. $3 - 6 \cdot 5 + 2$ **2.** $4^2 - 3^3 \cdot 3$ **3.** $-|-13| + |6 - (-3)|$

4. $9 + 25 \div 5 \cdot 3 + 2$ **5.** $3 - \left[12 \div (-3) + (4 \cdot 2 - 2)^2 \right]$

Solutions: 1. -25;   2. -65;   3. -4;   4. 26;   5. -29

Topic 3 - Some Basic Mathematical Definitions and Concepts

It is necessary to become familiar with some mathematical concepts which we use all the time in algebra. Recognizing these concepts and using them properly will lead to a deeper understanding of algebra and its problem solving capabilities.

Equivalence

Two numbers or groups of numbers that have the same quantitative value, even though they are expressed in different ways, are said to be equivalent.

Example 1. $-\dfrac{1}{2}$ is **equivalent** to -0.5 [$-\dfrac{1}{2}$ **has the same quantitative value as** -0.5 **even**

though the formats of the numbers are different]

Example 2. $2+3$ is **equivalent** to $4+1$ [$2+3$ **has the same quantitative value as** $4+1$
even

though the individual numbers are different.]

Example 3. 6 is equivalent to $2 \cdot 3$ [6 **has the same quantitative value as** $2 \cdot 3$. **Breaking down** 6 **into its equivalent,** $2 \cdot 3$, **is called** *factoring.*]

Interchangeable

Two numbers or groups of numbers that are equivalent are **interchangeable**. They can be **substituted for one another without changing quantitative value**. This becomes a useful tool when we are solving algebraic equations.

Example 1. $-\dfrac{1}{2}$ is **interchangeable** with. -0.5 **since their quantitative values**
are

the same. Therefore, $-\dfrac{1}{2}$ can be **substituted** for -0.5, in any mathematical situation.

Example 2. $2+3$ is **interchangeable** with $4+1$ **since their quantitative values are the same.** Therefore, $2+3$ can be **substituted** for $4+1$, or $4+1$ can be substituted for $2+3$. Can you think of any other different whole numbers that are interchangeable with $2+3$? [**5+0 is an example.**]

Example 3. $6x$ is **interchangeable** with $2x+4x$ **since their quantitative values are the same** . Even though we do not know the value of x, it must take on the **same value** in both expressions for a specific instance.

Example 4. $6x$ is **interchangeable** with $2x \cdot 3x$ **since their quantitative values are the same** . Even though we do not know the value of x, it must take on the **same value** in both expressions for a specific instance.

Absolute Value

As you know, all real numbers, except 0, can be either **positive** or **negative.** When we talk about the **absolute value of a number**, we look only at the magnitude [size] of the number without considering it's sign. **All absolute values result in a positive number**. The absolute value of a number is indicated by putting the number in between two **elongated vertical** [straight up and down] **lines.**

Examples: $|3|$ **or** $|-4|$

Since all absolute values are positive, **we can remove the absolute value symbol,** $|\ |$**, by making the value inside it positive.**

Examples: $|3| = 3$ **or** $|-4| = 4$

Another way of expressing this concept is thinking of absolute value as the **distance from the number to** 0 **on the number line**. In thinking about distances, **we realize that they must be positive**. For example, saying that the length of a pencil [a distance measure] is -4 inches makes no sense. A distance can't be less than 0. -4 is 4 units from 0 [from the left] on the number line. Therefore its absolute value is 4. 3 is 3 units from 0 [from the right] on the number line. Therefore its absolute value is 3.

We can **negate** an absolute value [make it negative] by inserting a negative sign **in front of** the absolute value symbol: **Examples:** $-|3| = -3$ **or** $-|-4| = -4$

The Commutative Property

If a mathermatical operation $[+, -, \times, \div]$ is **commutative** the order in which the numbers are connected by the operation does not affect its outcome. Let's look at these operations and see if "**order doesn't matter**".

Is addition commutative?

   Example: $3 + 4 = 7$ and $4 + 3 = 7$

   The order in which we connect 3 and 4 with addition **does not affect the result**. This would also be true of **any** real numbers connected by addition. Therefore,

The operation of addition is commutative.

Is subtraction commutative?

   Example: $3 - 4 = -1$ and $4 - 3 = 1$

   The order in which we connect 3 and 4 with subtraction **produces different results**. This would also be true of **any** real numbers connected by subtraction. Therefore,

The operation of subtraction is <u>not</u> commutative.

Is multiplication commutative?

Example: $3 \cdot 4 = 12$ and $4 \cdot 3 = 12$

The order in which we connect 3 and 4 with multiplication **does not affect the result**. This would also be true of **any** real numbers connected by multiplication. Therefore,

The operation of multiplication is commutative.

Is division commutative?

Example: $3 \div 4 = 0.75$ and $4 \div 3 = 1.\overline{3}$

The order in which we connect 3 and 4 with division **does affect the result**. This would also be true of **any** real numbers connected by division [with the exception of $1 \div 1$]. Therefore,

The operation of division is <u>not</u> commutative.

Changing an Integer to a Fraction

Any integer [zero, positive, or negative whole number] can be converted into a fraction of **equivalent** value by putting the integer over a denominator of 1. If you recall, a fraction is the division of the numerator by the denominator. When we put an integer over a denominator of 1, **we are dividing it by** 1. Dividing a number by 1 **does not change its value.**

Example: $\dfrac{5}{1}$ is **equivalent** to $5 \div 1$, which is **equivalent** to 5. So, $5 = \dfrac{5}{1}$

Equivalent Forms of a Negative Number

As you might recall, division of numbers with opposite signs **always results in a negative value.** The fraction, $\dfrac{-6}{2}$, can be viewed as the $-6 \div 2$. Since the dividend $[-6]$ has the opposite sign of the divisor $[2]$, the result must have a negative value, or -3. Consider the fraction $\dfrac{6}{-2}$. It also has a value of -3. Therefore, $\dfrac{-6}{2}$ is equivalent to $\dfrac{6}{-2}$ since they are both equivalent to -3. Putting this all together, we see that $\dfrac{-6}{2} = \dfrac{6}{-2} = -3$. All three of these numbers are **interchangeable!**

Another example of equivalent and interchangeable numbers to -3 is as follows:

$-3 = \dfrac{-3}{1} = \dfrac{3}{-1}$. **[They are equivalent and interchangeable since they all have the quantitative value of -3 .]**

It is important to notice that $\dfrac{-3}{1} = \dfrac{3}{-1}$ is also **interchangeable** with $-\dfrac{3}{1}$, since they all

have a **negative** value .

Reciprocals

The reciprocal of any number can be found by:

      (1) changing the specified number to an equivalent fraction **[if it is not already a fraction]**, and

      (2) reversing the numerator and denominator of the fraction.

Examples:	Number	The Number Changed to an Equivalent Fraction	The Number's Reciprocal
	4	$\dfrac{4}{1}$	$\dfrac{1}{4}$
	-3	$\dfrac{-3}{1}$ or $-\dfrac{3}{1}$	$\dfrac{1}{-3}$ or $-\dfrac{1}{3}$

These are basic tools that **we use all the time in algebra** when "**operating**" mathematically **[adding, subtracting, multiplying, dividing]**. The following related concepts are also **fundamentally important for gaining a true understanding of algebra.**

Identity Elements

An **identity element** is a number that, when used in an *operation*, **[addition, subtraction, multiplication or division]**, **does not change the value of the number with which it is "operating".** Identity elements are **not necessarily** the same for different operations.

For example, the identity element for **addition and subtraction** is the same , but **multiplication and division** have a **different** identity element from addition and subtraction. Let's start by looking at the **addition and subtraction operations.**

1. The Identity Element for Addition and Subtraction

The following examples will help understand the idea:

    The Addition Operation

        Example 1. $3 + 0 = 3$

In this case we **added [operation]** 0 to the number 3 which resulted in 3, **the number with which we started.** Therefore, "0" is **the identity element for addition** for 3.

Example 2. $-5 + 0 = -5$

In this case we **added** [operation] 0 to the number -5 which resulted in -5, the number with which we started. Therefore, "0" is the **identity element for addition** for -5.

In fact, think about adding zero to *any* real number. The result will be the original real number!

Therefore 0 is the <u>identity element</u> for <u>addition</u> for *all* real numbers.

The Subtraction Operation:

Example 1. $4 - 0 = 4$

In this case we **subtracted** [operation] 0 from the number 4 which resulted in 4, the number with which we started. Therefore, "0" is the identity element for subtraction for 4.

Example 2. $-6 - 0 = -6$

In this case we **subtracted** [operation] 0 from the number -6 which resulted in -6, the number with which we started. Therefore, "0" is the identity element for subtraction for -6.

In fact, if we subtract zero from *any* real number, the result will be the original real number!

Therefore 0 is the <u>identity element</u> for <u>subtraction</u> for *all* real numbers.

2. Identity Elements for Multiplication and Division

We will stay with this idea of operating on a number and getting the same number back, but this time we will consider **the multiplication and division operations.** Look at the following examples:

Examples: (1) $2 \cdot 1 = 2$ [The **operation** here is **multiplication**.]

(2) $2 \div 1 = 2$ [The **operation** here is **division**.]

(3) $\dfrac{2}{1} = 2$ [The **operation** here is **division** in the form of a fraction.]

(4) $-5 \cdot 1 = -5$ [The **operation** here is **multiplication**.]

(5) $\dfrac{-5}{1} = -5$ [The **operation** here is **division** in the form of a **fraction**.]

In fact, multiplying or dividing by the number 1 will always give you back the number with which you started. Therefore 1 is the <u>identity element</u> for <u>multiplication or division</u> for *all* real numbers.

Inverses

A concept that is closely related to this idea of an **identity element** is the concept of **inverse.** A number that is operated [added, subtracted, multiplied or divided] on with its **inverse** results in the **identity element** for that operation.

This idea can be "modeled" in the following way:

Original Number [operation $(+,-,\times,\div)$] **Inverse = identity element** [for that operation]

1. Inverses for addition [also called opposites]

Inverses [also called "opposites"], when **added** to each other [the addition operation], **always result in zero**, which is the **identity element for addition!**

Example 1.

3	+	−3	=	0
original number	operation	inverse for addition		identity element for addition

Example 2.

−5	+	5	=	0

Example 3.

999	+	−999	=	0

The pattern with **opposite numbers becomes evident.** [They are numbers with the same *absolute value*, but with opposite signs.] When added, they will always result in 0 [The identity element for addition].

The **opposite** of a number is technically called its **additive inverse**. An operation [in this case, addition] that results in the **identity element** [in this case "0"] always involves some number and its **inverse**.

2. Inverses for Subtraction.

We use the same model as we did for finding additive inverses, only this time the **operation is subtraction**:

Original Number − **inverse** = **identity element [for subtraction]**

Example 1.

8	−	?	=	0
original number		inverse for subtraction		identity element for subtraction

If we subtract 8 from itself, we get 0:

8	−	8	=	0

So the inverse for 8 under the subtraction operation is itself (8).

Example 2: $\quad\quad$ -5 $\quad$ $-$ $\quad$? $\quad$ = $\quad$ 0

original number $\quad\quad$ inverse $\quad\quad$ identity element [for subtraction]

$$-5 \quad - \quad (-5) \quad = \quad 0$$

Changing subtraction to addition: $\quad -5 \quad + \quad 5 \quad = \quad 0$

So, the **inverse under subtraction** for -5 is -5 **[the original number !]**.

Notice that **identity elements** are **always the same for an operation**, regardless of the number with which we start. **Inverses**, however, **differ** depending upon the number we start with.

The inverse under subtraction for any real number is the number itself.

3. Inverses for Multiplication.

When we select any real number, we look for another number as a multiplier **[the multiplication operation]** such that, when these numbers are multiplied, the result is the number 1 **[the identity element for multiplication]**.

This can be "modeled" in the same ways as we did for addition and subtraction, except that the operation in this case is multiplication:

Original number $\quad \times \quad$ **inverse** $\quad = \quad$ **the identity element for multiplication [1]**

Example 1. What is the inverse of 5 under the **multiplication** operation?

Using our model: $\quad\quad$ 5 $\quad\quad$ $\cdot$ $\quad\quad\quad$? $\quad\quad\quad$ = $\quad\quad\quad$ 1

$\quad\quad\quad\quad$ original number $\quad\quad$ inverse for multiplication $\quad\quad$ identity element for multiplication

Let's try **the reciprocal** of 5 and see if it makes the model true:

$$5 \quad \cdot \quad \frac{1}{5} \quad = \quad 1$$

Is this a true statement? It is easier to see if we change the 5 to an **equivalent** fraction:

$$\frac{5}{1} \quad \cdot \quad \frac{1}{5} \quad = \quad 1$$

Multiplying the fractions results in:

$$\frac{5 \cdot 1}{1 \cdot 5} \quad = \quad \frac{5}{5} \quad = \quad 1$$

This is a true statement! So the inverse of 5 under multiplication [called the **multiplicative inverse**] is $\dfrac{1}{5}$ [another name for reciprocal]

Example 2. What is the multiplicative inverse of -6 [under the multiplication operation]?

Let's change -6 to its equivalent fraction:

$$\underset{\text{original number}}{\dfrac{-6}{1}} \quad \cdot \quad \underset{\text{inverse for multiplication}}{?} \quad = \quad \underset{\text{identity element for multiplication}}{1}$$

Multiplying $\dfrac{-6}{1}$ by its reciprocal: $\qquad \dfrac{-6}{1} \cdot \dfrac{1}{-6} = 1$

Multiplying the fractions results in: $\qquad \dfrac{-6 \cdot 1}{1 \cdot (-6)} = \dfrac{-6}{-6} = 1$

This is a true statement! So the multiplicative inverse of -6 is its reciprocal, $\dfrac{1}{-6}$ or its equivalent, $-\dfrac{1}{6}$.

Also notice that the **reciprocal of a negative number is also negative**!

As can be seen, each inverse for multiplication changes as the "original number" changes. **BUT** they all have something in common with the original number. The inverse is always the original number's **reciprocal.**

> **The reciprocal [multiplicative inverse] of a number is its inverse under the multiplication operation.**

4. Inverses for Division.

The inverse for division can be "modeled" in the same ways as we did for multiplication, except that the operation in this case is division:

Original number $\div$ inverse $=$ the identity element for division [1]

Example 1. What is the inverse of 5 under the division operation?

Using our model: $\qquad \underset{\text{original number}}{5} \quad \div \quad \underset{\text{inverse for division}}{?} \quad = \quad \underset{\text{identity element for division}}{1}$

The only number that will divide 5 and result in 1 is 5.

$$5 \quad \div \quad 5 \quad = \quad 1$$

Or putting the division in fractional form: $\quad \dfrac{5}{5} \; = \; 1$

Example 2. What is the inverse of -6 [under the division operation]?

$$\underset{\text{original number}}{-6} \quad \div \quad \underset{\text{inverse for division}}{?} \quad = \quad \underset{\text{identity element for division}}{1}$$

The only number that will divide -6 and result in 1 is -6.

$$-6 \quad \div \quad -6 \quad = \quad 1$$

Or putting the division in fractional form: $\quad \dfrac{-6}{-6} \; = \; 1$

We can see the pattern that is emerging:

The inverse of a number under the division operation is the number itself.

To **summarize**, we see that **identity elements** and **inverses** *have a connection:*

- **"0" is the identity element for addition and subtraction.**

- **Adding a number and its *opposite* [additive inverse] results in 0** [the identity element for addition].

- **Subtracting a number from *itself* [the inverse for subtraction] results in 0** [the identity element for subtraction].

- **"1" is the identity element for multiplication and division.**

- **Multiplying a number by its *reciprocal* [multiplicative inverse] results in "1"** [The identity element for multiplication]

- **Dividing a number by *itself* [the inverse for division] results in "1"** [the identity element for division].

These are very important concepts that are *utilized constantly in algebra.*

A Concluding Thought

Why are these ideas so important for understanding and utilizing algebra?

These concepts contain the logical underpinnings for arithmetic as well as algebra and are used constantly in solving algebraic equations.

We often have to convert numbers or algebraic expressions to some **equivalent** form [having the same quantitative value] using **identity elements** since **identity elements don't change their value.** We often add an **opposite** [additive inverse] to both sides of an equation to get the variable alone on one side of the equation. We also multiply by **reciprocals** [multiplicative inverses] to change **coefficients** to "1" [identity element for multiplication]. We will define the terms, algebraic expressions, equations, and coefficients shortly so these ideas will become more concrete.

As we start to manipulate algebraic equations to obtain a solution to a problem, it is important to understand the **logic** behind these manipulations. Then algebra becomes **comprehensible** as a consistent structure that will always yield the correct result and will then be as meaningful as the arithmetic structure with which we are already familiar.

If we just memorize algebraic procedures without understanding their validity, then it becomes difficult to utilize this system in unfamiliar situations, which would greatly limit this wonderful mathematical tool.

Concept Homework

As an assessment of your understanding of the concepts set forth in this section, answer the following questions. If necessary, review the material to help you arrive at the correct conclusions. With "true or false" statements, if the statement is true, make up an example that supports the statement without using examples given in the material. If a statement is false, correct the wording to make it true. Then make up an example that supports the statement without using examples given in the material.

True or False:

a. Two numbers, or groups of numbers [connected by an operation symbol], are *equivalent* if they have the same quantitative value.

b. When we operate $[+,-,\times,\div]$ on a number with an identity element, the result is always that number with which we started.

c. Identity elements can take on different forms as long as they have an equivalent quantitative value. [For example, 1 and $\frac{3}{3}$ are both the identity element for multiplication.]

d. The identity element for the multiplication and division operations is 0.

e. The inverse under the addition operation for -3 in addition is **itself** $[-3]$.

f. The inverse for the addition operation is always constant, regardless of the number with which we start.

g. In the division operation, a number and its inverse are the same.
[Use # ÷ inverse = 1 **and substitute any number for # to verify whether this question is true or false.**]

h. Another name for " multiplicative inverse" is "reciprocal".

i. For any operation: "some number" $[+,-,\times,\text{or }\div]$ identity element $=$ inverse.

j. "Some number" ÷ "the same number" = 1. Therefore, the inverse of a number is the number itself when the operation is division.

Solutions: a. True; b. True; c. True; d. False; e. False; f. False; g. True; h.t rue; i. False; j. True.

Exercises

1. Replace the question marks in the following equations with a number that will make them true:

   a. $3 + ? = 3$ **b.** $-6 + ? = -6$ **c.** $5 - ? = 5$ **d.** $-2 - ? = -2$

2. What is the identity element for addition and subtraction?

3. Replace the question marks in the following equations with a number that will make them true:

   a. $8 \cdot ? = 8$ **b.** $-9 \cdot ? = -9$ **c.** $11 \div ? = 11$ **d.** $\dfrac{3}{?} = 3$

4. What is the identity element for multiplication and division?

5. What is another name for the additive inverse?

6. Find the inverse under addition for the following numbers?

   a. 8   **b.** -15 **c.** 25   **d.** -393

7. Find the inverse under subtraction for the following numbers?

   a. 8   **b.** -15 **c.** 25   **d.** -393

8. Find the inverse under multiplication [multiplicative inverse] for the following numbers?

   a. 8   **b.** -15 **c.** 25   **d.** -393

9. Under the multiplication operation, how is a number and its inverse related?

10. Find the inverse under division for the following numbers?

    a. 8   **b.** -15 **c.** 25   **d.** -393

11. Under the division operation, what is the relationship between a number and its inverse?

Solutions: 1a. 0; 1b. 0; 1c. 0; 1d. 0; 2. 0; 3a. 1; 3b. 1; 3c. 1; 3d. 1; 4. 1; 5. opposite; 6a. -8; 6b. 15; 6c. -25; 6d. 393;

7a. 8; 7b. -15; 7c. 25; 7d. -393; 8a. $\dfrac{1}{8}$; 8b. $-\dfrac{1}{15}$; 8c. $\dfrac{1}{25}$; 8d. $-\dfrac{1}{393}$; 9. They are reciprocals of each other.

10a. 8; 10b. -15; 10c. 25; 10d. -393; 11. They are the same number.

Topic 4 - Fractions and the Identity Element

Understanding fractions and how we operate [+,−,×,÷] with them is crucial for our manipulations in algebra as they occur frequently in algebraic equations. We will therefore review these operations and look at how these operations depend upon utilizing identity elements. There are many ways of thinking about fractions. We have already considered that fractions can be understood as a representation of a division problem with the **divisor** [**the number doing the dividing**] as denominator [**bottom of fraction**] and the **dividend** [**the number being divided**] as numerator [**top of fraction**].

$$\textbf{Example:} \quad 1 \div 3 \Rightarrow \frac{1}{3}$$

We have also learned in our study of the signed number rules for division that if the **signs of the numerator and denominator are the same**, the value of these **fractions must be positive**.

$\frac{1}{3}$ and $\frac{-1}{-3}$ are equivalent as they both have the positive value of $\frac{1}{3}$. So, $\frac{1}{3}$ and $\frac{-1}{-3}$ are **interchangeable**.

If we consider the case where the signs in the numerator and denominator are opposites, our division rule tells us that the results **will always** be **negative**:

$\frac{1}{-3}$ and $\frac{-1}{3}$ are **equivalent** as they both have the negative value of $-\frac{1}{3}$. So,

$\frac{1}{-3}$ and $\frac{-1}{3}$ and $-\frac{1}{3}$ are all **interchangeable**.

Here is another way to think about fractions:

The denominator divides "the whole", [**we can think of the whole as the number "1" or "1 of something"**] into some **number of equal parts.** The number value of the denominator expresses this division of "the whole" into equal parts. The **numerator** represents **how many of these equal parts are contained in the fraction.** For example, with the fraction, $\frac{1}{3}$, the value of this number would be **"1 of 3 equal parts of the whole"**. [**In this case "the whole" is divided into three equal parts.**]

Other examples: $\quad \frac{2}{7} \rightarrow$ two of seven equal parts [**the whole being divided into 7 equal parts**].

$$\frac{3}{10} \rightarrow \text{3 of 10 equal parts.'' [\textbf{the whole being divided into 10 equal parts}].}$$

Multiplication of Fractions

When we multiply fractions, we are merely *combining factors*. Factors are numbers that are multiplying each other. **[We will discuss factors and factoring at length later in the book.]**

Example: Consider the multiplication of $\frac{2}{3}$ by $\frac{3}{4}$: $\quad \frac{2}{3} \cdot \frac{3}{4} \Rightarrow \frac{2 \cdot 3}{3 \cdot 4}$.

We can multiply these **factors**, resulting in: $\quad \frac{2 \cdot 3}{3 \cdot 4} = \frac{6}{12}$ [one solution].

Hence, the rule for fractional multiplication is:

When multiplying fractions, we multiply both the numerators and denominators.

Instead of immediately doing this multiplication, it is often more convenient to just leave the factors as they are without performing the multiplication: $\quad \frac{2}{3} \cdot \frac{3}{4} = \frac{2 \cdot 3}{3 \cdot 4}$

Now, let's **re-order** the **factors** in the denominator: $\quad = \frac{2 \cdot 3}{4 \cdot 3}$

Note: We are "allowed" to re-order factors because multiplication is *commutative*. In other words, order doesn't matter when multiplying! **[Example: $3 \cdot 4$ will yield the same result as $4 \cdot 3$]**

We know that $\frac{3}{3}$ has a value of "1" $[3 \div 3]$, the **identity element**

For multiplication!

So we are really multiplying $\frac{2}{4}$ by the equivalent of 1: $\quad = \frac{2}{4} \cdot 1$

Multiplying by 1 **[the identity element for multiplication]**

does not change the quantitative value of $\frac{2}{4}$: $\quad \frac{2}{4} \cdot 1 = \frac{2}{4}$

That is why we can **"cancel"** it out!

Therefore: $\quad \frac{2 \cdot \cancel{3}}{4 \cdot \cancel{3}} = \frac{2}{4}$

[solution]

Since $\frac{2 \cdot 3}{3 \cdot 4} = \frac{6}{12}$ and $\frac{2 \cdot 3}{4 \cdot 3} = \frac{2}{4}$, $\frac{6}{12}$ must be equivalent to $\frac{2}{4}$

[Two values equal to the same value must be equal to each other!]

Further **Note: Back in the day, we were taught that we could "cancel" the 3's in the above problem. What this *really* means is that we are *eliminating the identity elements* in this fractional multiplication, since identity elements do not change values.**

You might be thinking that the solution $\frac{2}{4}$ can be "reduced" to $\frac{1}{2}$. This is *quite true*. This reduction process is *also* "eliminating identity elements" as can be seen below:

Let's **factor** [breakdown into multipliers that have the same value] $\frac{2}{4}$: $\frac{2}{4} = \frac{2 \cdot 1}{2 \cdot 2}$

Eliminating or "canceling" the identity element, $\frac{2}{2}$,[another form of the number 1]: $\frac{\cancel{2} \cdot 1}{\cancel{2} \cdot 2} = \frac{1}{2}$

We were taught that to "*reduce* a fraction, we divide " top and bottom" by 2. In the above example, removing the identity element is **equivalent** to dividing top and bottom by 2. The problem was that we were never taught **why** we could do this without changing the fraction's value! Dividing top and bottom of a fraction by the same number, is actually removing an **identity element, 1, for division! Dividing a fraction by 1 doesn't change its quantitative value**. Therefore, "reducing" a fraction doesn't change its quantitative value.

Let's go back to our original example: $\frac{2}{3} \cdot \frac{3}{4}$

How would we entirely "reduce" this fraction by eliminating identity elements? Let's start by finding all the *prime factors* the fraction. These are all the factors that cannot be **further factored** except by the factor 1. The numerator is already broken down into prime factors.

We will now do the same with the denominator: $\frac{2}{3} \cdot \frac{3}{4} = \frac{2 \cdot 3 \cdot 1}{3 \cdot 2 \cdot 2}$

We can now use the commutative property to re-order the factors: $\frac{2 \cdot 3 \cdot 1}{3 \cdot 2 \cdot 2} = \frac{2 \cdot 3 \cdot 1}{2 \cdot 3 \cdot 2}$

We will now "cancel"the identitiy elements from the fraction: $\frac{\cancel{2} \cdot \cancel{3} \cdot 1}{\cancel{2} \cdot \cancel{3} \cdot 2} = 1 \cdot 1 \cdot \frac{1}{2}$

Resulting in the totally reduced fraction: $\frac{1}{2}$

[solution]

This idea is incorporated in the general rule:

$$\frac{a \cdot c}{b \cdot c} = \frac{a}{b} \quad \text{[where } b \text{ and } c \text{ are non-zero numbers.]}$$

Division of Fractions

Division can be defined as multiplication by the **reciprocal** of the divisor. Therefore, **we can change any division problem to an equivalent multiplication problem by substituting the divisor with its reciprocal.**

Example: Divide: $\dfrac{3}{4} \div \dfrac{1}{2}$

We **change the division to multiplication** by taking the reciprocal of $\dfrac{1}{2}$ [the divisor]: $\dfrac{3}{4} \cdot \dfrac{2}{1}$

Combining the two fractions into one

[the same as multiplying the numerators and denominators]: $\dfrac{3}{4} \cdot \dfrac{2}{1} = \dfrac{3 \cdot 2}{4 \cdot 1}$

Instead of just "canceling" by dividing both the 4 in the denominator and

the 2 in the numerator both by 2 , let's factor the 4 in the denominator: $\dfrac{3 \cdot 2}{4 \cdot 1} = \dfrac{3 \cdot 2}{2 \cdot 2 \cdot 1}$

$\dfrac{2}{2}$ is an identity element so it can be removed: $\dfrac{3 \cdot \cancel{2}}{2 \cdot \cancel{2} \cdot 1} = \dfrac{3}{2 \cdot 1} = \dfrac{3}{2}$

So: $\dfrac{3}{4} \div \dfrac{1}{2} = \dfrac{3}{2}$

Addition of Fractions

Suppose we wanted to add two fractions together. Let's add: $\dfrac{2}{6} + \dfrac{3}{6}$.

We would then have "2 of 6 equal parts of the whole" plus an additional "3 of 6 equal parts of the whole" for a total of "5 of 6 equal

parts of the whole". Mathematically, we would express this as: $\dfrac{2}{6} + \dfrac{3}{6} \Rightarrow \dfrac{2+3}{6} = \dfrac{5}{6}$

When seeing this problem presented mathematically, we would be tempted to add **both** the numerators **and** the denominators resulting in $\dfrac{5}{12}$. But when we think of a fraction in the way suggested above, we see that the magnitude of the fraction is represented by the **numerators** with the denominators telling us how many equal parts there are in "the whole". **Adding the fractions above didn't change how many equal parts there were in the whole [6],** so there is no reason to add the denominators together. Hence, the rule for adding fractions:

When fractions have the same denominator, only the numerators get added.

[Note: The addition rule for fractions is quite different from the multiplication rule. Now we can understand*hy.*]

This rule only holds when the denominators are the same. That's not always the case. **"What if the denominators are different?"** In other words, **"What if the fractions being added, divide 'the whole' differently?"**

Example.   Add: $\dfrac{2}{3} + \dfrac{3}{4}$.

We just can't add the numerators because the first fraction is "2 of 3 equal parts of the whole" and the second is "3 of 4 equal parts of another whole". "The whole" is not divided the same way with each fraction. How do we solve this problem?

We **transform** each fraction into an ***equivalent fraction*** [one that has different numbers but the same quantitative value.] where the denominators for both are the same! And once we do, we can then add the resulting fractions, using our rule for the addition of fractions. This is the idea behind **"finding common denominators"**. This process of transforming fractions to **equivalent** forms with the same denominators is really utilizing **the identity element for multiplication and its properties.** This example will illustrate this point.

We must find a common denominator in order to add these fractions **while at the same time not changing the value of the fractions**. We can use two **identity elements in different**
forms that both have a value of "1" to accomplish this! Remember that multiplying by the identity element, 1, does not change the value of a number. The good news is that 1

comes in many equivalent forms: for example, 1 is equivalent to $\dfrac{3}{3}$ and also $\dfrac{4}{4}$.

Lets multiply each fraction by these ***convenient* identity elements** that

suit our objective of **creating common denominators:** $\dfrac{2}{3} \cdot \dfrac{4}{4} + \dfrac{3}{4} \cdot \dfrac{3}{3}$

We have **not** changed the original value of the fractions, $\dfrac{2}{3} + \dfrac{3}{4}$ since

each fraction has been multiplied by the equivalent of 1. Performing the

multiplication with each group of fractions: $\dfrac{2 \cdot 4}{3 \cdot 4} + \dfrac{3 \cdot 3}{4 \cdot 3}$

Results in: $\dfrac{8}{12} + \dfrac{9}{12}$

We have created common denominators that allow us to add the fractions without having changed the value of original fractions! We

can now add the numerators: $\dfrac{8 + 9}{12} = \dfrac{17}{12}$

So: $\dfrac{2}{3} + \dfrac{3}{4} = \dfrac{17}{12}$

[solution]

[Note: $\dfrac{17}{12}$ is 17 of 12 equal parts of the whole. This indicates that the value of the fraction is

greater than 1 which is equivalent to $\dfrac{12}{12}$. Therefore, it falls into the category of "**improper fractions**".]

Improper fractions can be changed into *mixed numbers* if we wish. For example, $\frac{17}{12}$ can be thought of as $\frac{12}{12} + \frac{5}{12}$ which is equivalent to $1\frac{5}{12}$ [which actually means $1 + \frac{5}{12}$]. The advantage of converting to mixed numbers is that they have more meaning in the real world. For example, if we are measuring a distance in miles, we have an idea of the length of $1\frac{5}{12}$ miles [a little less than $1\frac{1}{2}$ miles which equals $1\frac{6}{12}$ miles]. This has more meaning for us than merely saying $\frac{17}{12}$ miles.

The disadvantage of mixed numbers is that they are hard to **operate** $[+, -, \times, \div]$ with. For example, we would have an easier time multiplying $\frac{17}{12}$ by $\frac{2}{3}$ $\Rightarrow \left[\frac{17}{12} \cdot \frac{2}{3} \right]$ than multiplying $1\frac{5}{12}$ by $\frac{2}{3}$ $\Rightarrow \left[1 \cdot \frac{2}{3} + \frac{5}{12} \cdot \frac{2}{3} \right]$.

Concept Homework

As an assessment of your understanding of the concepts set forth in this section, answer the following questions. If necessary, review the material to help you arrive at the correct conclusions.

1. $\frac{-1}{-2} = \frac{1}{2}$. Is this statement true? Explain why or why not.

2. Consider $\frac{-1}{2}$, $\frac{1}{-2}$, $-\frac{1}{2}$. Are these fractions interchangeable? Explain why or why not.

3. Show an example of how we eliminate identity elements in order to reduce fractions.

4. Explain the relationship between division and multiplication. Then, give an example of a division problem that is equivalent to a corresponding multiplication problem.

5. Explain why don't we add denominators when adding fractions.

6. Comment on the following statement:

   "Adding fractions with different denominators is like trying to add apples and oranges".

7. Consider the following multiplication problems:

   a. $3\frac{3}{8} \cdot \frac{1}{3}$ and   **b.** $\frac{27}{8} \cdot \frac{1}{3}$

   Are these problems equivalent?
   Perform both multiplications.
   Which do you think was less complicated?

Exercises

1. Reduce the following fractions to lowest terms by factoring out identity elements:

a. $\dfrac{4}{12}$ b. $\dfrac{21}{35}$ c. $\dfrac{35}{28}$ d. $\dfrac{90}{125}$

2. Multiply and simplify. Use factoring techniques to eliminate identity elements:

a. $\dfrac{9}{4}\cdot\dfrac{16}{15}$ b. $\dfrac{7}{8}\cdot\dfrac{12}{28}$ c. $\dfrac{10}{16}\cdot\dfrac{16}{9}\cdot\dfrac{3}{5}$ d. $\dfrac{7}{8}\cdot\dfrac{4}{21}\cdot\dfrac{2}{3}$

3. Divide and simplify. Use factoring techniques to eliminate identity elements:

a. $\dfrac{22}{7}\div\dfrac{15}{14}$ b. $\dfrac{32}{39}\div\dfrac{64}{51}\div\dfrac{17}{52}$

4. Perform the following operations and reduce if necessary:

a. $\dfrac{1}{6}+\dfrac{4}{6}-\dfrac{2}{6}$ b. $\dfrac{1}{2}+\dfrac{1}{3}$ c. $\dfrac{1}{4}+\dfrac{3}{8}-\dfrac{5}{12}$

5. Perform the indicated operations and reduce [keep in mind the rules of "order of operations}:

a. $\dfrac{2}{3}+\dfrac{7}{8}\cdot\dfrac{4}{14}-\dfrac{1}{6}$ b. $\dfrac{23}{24}-\dfrac{7}{12}\div\left(\dfrac{35}{36}-\dfrac{7}{9}\right)$

Solutions: 1a. $\dfrac{1}{3}$; 1b. $\dfrac{3}{5}$; 1c. $\dfrac{5}{4}$; 1d. $\dfrac{18}{25}$; 2a. $\dfrac{12}{5}$ or $2\dfrac{2}{5}$; 2b. $\dfrac{3}{8}$; 2c. $\dfrac{2}{3}$; 2d. $\dfrac{1}{9}$; 3a. $\dfrac{44}{15}$ or $2\dfrac{14}{15}$; 3b. 2; 4a. $\dfrac{1}{2}$; 4b. $\dfrac{5}{6}$; 4c. $\dfrac{5}{24}$; 5a. $\dfrac{3}{4}$; 5b. $-\dfrac{49}{24}$

Topic 5 - Some Definitions for Algebra

Knowing the basic vocabulary of algebra is important for understanding since these terms are used constantly as we build algebraic ideas.

Variable

The concept of variable can take on many different meanings depending upon the situation. For the purposes of this course, we will consider **three** of these meanings:

1. A variable can represent a unique **[one and only one]** unknown number.
   Example: $x + 3 = 6$ ["x", in this case, represents a unique unknown number which will **make the statement true**. In this case it is easily determined be 3.]

Notice that there is only one variable in the equation and only one possible value for x that will make the equation true.

2. A variable can represent two different possible values if it is raised to the 2nd power.
   Example: $x^2 + 3 = 7$ [x, in this case, could represent either 2 or -2, since when we square, or raise to the second power, either of these values, it results in 4 which would **make the statement true**. We will define **powers** or **exponentials** later in this topic.]

3. A variable can represent many different values depending upon other variables contained in an equation. This will happen when there are **more than one variable in an equation**.
   Example: $A = lw$.

You may recognize this as the **formula** [Formulas are equations that always have more than one variable.] It represents the **relationship** between the area, the length and width of a rectangle. As we change the values of "l" [representing the length of a rectangle] and/or "w" [representing the width of a rectangle], the value of A [representing the area of the rectangle] will also change. Therefore, the value of A **depends** on the values of l and w. Although A keeps changing as l and/or w changes, the *relationship* between A, l, and w is always the same [A will always equal l times w.]. A can take on an **infinite** number of values since l and w can take on an **infinite** number of values.

Terms & Algebraic Expressions

For our purposes, we will define a **term** as a number, **or** a variable, **or** a combination of a number and variable(s) connected by the **multiplication or division operations**.

When we combine terms with **addition or subtraction operations**, they become an **expression**. If an expression contains a variable it is considered to be an **algebraic expression**. **Examples** of **terms**: 3; x; $4y$; $\dfrac{y}{3}$; $\dfrac{4}{5}a$; $\dfrac{abc}{d}$

42

Examples of **expressions**: $3+4$; $2+(3-2)^2$

Examples of **algebraic expressions**: $3y+1$; $\dfrac{p}{2}+2x$; $3a^2-4b+5c^3-\dfrac{5}{4}d$

Coefficients and Constants

These words are both defined as **numbers** as opposed to **variables**. However, they use numbers in different ways:

A **coefficient** is a number used to **multiply a variable**, while a **constant** is a **stand-alone number** in an algebraic expression.

> **Example**: In the algebraic expression $3x+2$, the 3 is the **coefficient** of x [It multiplies x], while the 2 is a **constant** [It stands alone in the algebraic expression.].

If a variable, doesn't have a coefficient, [for example "x" or "y"], there is an implied coefficient of 1. ["$1x$" or "$1y$"]. This is true because **multiplying by 1 doesn't change value. 1 is the identity element for multiplication!** At times, it is convenient to express a variable [x for example] as $1x$ to clarify the fact that the variable has the **implied** coefficient of 1.

A Factor of a Number

A factor of a number is a **whole** number that will divide that number evenly.

> **Examples:** **(1)** 1, 2, 3, 4, 6 and 12 are factors of 12. **[They all divide 12 evenly.]**
>
> To look at it another way, 1, 2, 3, 4, 6 and 12 [factors of 12] each can be *multiplied* by another whole number to result in 12.
> [$1\cdot12=12$; $2\cdot6=12$; $3\cdot4=12$]
>
> **(2)** 7 and 1 are factors of 7 **[They both divide 7 evenly and $7\cdot1=7$.]**
>
> **(3)** Variables can also have factors. p *is a factor of* p^2.
> **[It divides p^2 evenly. $p\cdot p=p^2$]**

We will discuss **exponentials** [numbers or variables raised to a power] below, and later in the book more thoroughly.

To **factor a number**, means to break it down into numbers that, when multiplied, have the **same quantitative value** as the original number.

> **Examples:** **(1)** 12 factors into $3\cdot4$ which is **equivalent** and **interchangeable** with 12.
> **(2)** 12 also factors into $2\cdot2\cdot3$ [these are *prime* factors.] which are **equivalent** and **interchangeable** with 12.

A **prime number** is a number whose only factors are itself and 1.

Examples: **(1)** 11 is a prime number because its **only** factors are 11 and 1.
(2) 2 is a prime number because its **only** factors are 2 and 1.

Exponential Expressions

x^a is the general form of an *exponential expression*. In this form, x is called the base and a is called the exponent. a determines **how many times** x **is used as a factor**. If there is **no exponent**, we assume that the exponent is 1. For example, y is **equivalent** to and **interchangeable** with y^1 when expressed as an exponential expression. **[By the above definition, y^1 is y taken as a factor once, so $y^1 = y$.]**

Examples: $x^3 = x \cdot x \cdot x$ or $3^4 = 3 \cdot 3 \cdot 3 \cdot 3$ **[which has a quantitative value of 81.]**

Further Example: $(-1)^3 = -1$.

[$(-1)^3$ is defined as -1 taken as a factor three times. Therefore, $(-1)^3 = (-1)(-1)(-1)$. Multiplying this out results in -1.]

Simplifying Algebraic Expressions

We can use **addition, subtraction** or the **distributive property** to simplify algebraic expressions.

When using the **addition and subtraction operations** to simplify, we **combine " like terms"**. Only **like terms can be added or subtracted.**

Like terms are such that **all variables in each term** [and their exponents], are *exactly* the same. Only the **coefficients** can differ in like terms. **Only the coefficients are added or subtracted. The variable parts stay as is.**

Examples: (1) $3x^2 + 2x^2 = 5x^2$ **[In this example, "3", "2" and "5"** are coefficients and $3x^2$ and $2x^2$ are like terms.**]**

A good way to think of like terms such as $3x^2 + 2x^2$ **is as follows: There are 3 of "these"** $[x^2]$ **and 2 more** of **"these"** $[x^2]$. So now we have 5 of **"these"** $[x^2]$ or $5x^2$. Notice that the variable parts **["these"]** must be exactly alike or they would not be **"these"**. When the terms are combined, the variables **stay the same** and the **coefficients** [in this case "3" and "2"] are **added**.

(2) $4a^2bc - 6a^2bc \Rightarrow 4a^2bc + \left(-6a^2bc\right)$ [We changed subtraction to addition.]

$\Rightarrow -2a^2bc$ [We added like term's coefficients and left the variable parts as is.]

(3) $5a^3bc^2d + 6a^3b^2cd$ **cannot be combined**, since the variable parts of the terms are not **exactly** the same. **Some of the variables have different exponents.**

Note: $2ab + 3ba$ **are** like terms! The order in which the variables are listed does **not** affect their quantitative value since order doesn't matter under the multiplication operation. this is an example of the **commutative property of multiplication**. Would you think the addition operation is also commutative? [yes]. How about the subtraction operation? [no]

Simplifying Using the Distributive Property

In the case, $3(a + b)$, we **distribute** this expression by multiplying 3 by **every** term in the parenthesis:

Examples: **(1)** $3(a + b) \Rightarrow 3(1a + 1b) = 3a + 3b$

[Inside the parenthesis, we first expressed the "hidden" or implied coefficient of a and b. We did this step before we distributed.]

(2) $-5(x - 4) \Rightarrow -5(1x - 4) = -5x + 20$

This next case is very important to understand as it often causes confusion when manipulating algebraic equations.

(3) $2x - (3x + 2y - 4) \Rightarrow 2x + (-1)(3x + 2y - 4)$

[To change a subtraction to addition before *a parenthesis* we must insert a (-1) multiplier directly in front of the parenthesis.]

$\Rightarrow 2x + (-3x) - 2y + 4$

[The multiplication by (-1) results in *reversing the sign* of *every* term within the parenthesis!]

$\Rightarrow -1x - 2y + 4$ [Combining like terms.]

The result of changing the subtraction operation **in front of the parenthesis** to addition is that *every* term in the parenthesis has its sign reversed!

In general, the **distributive property** can be represented by :

$a(b + b) = ab + ac$ where a, b and c represent real numbers.

The Concept of Equations as Equivalent Values

Every equation contains an "equals" $[=]$ symbol. If the equation is a true statement, then the total value of the terms or expressions that are on the left side of the equal sign, **by definition**, has the same quantitative value as the total value of the terms or expressions that are on the right side of the equal sign. The equation is like a seesaw in perfect equilibrium with the equal sign as the fulcrum point. The seesaw is balanced when there are equal "weights" [**quantitative values**] on each side. If an equation is a true statement, it has to be in balance with the right side **equivalent** to the left side.

When we work with equations **we must always keep this balance** or they will no longer be true and therefore will be worthless to us. Therefore, if we **add or subtract** some quantity to/from one side, we must also add or subtract **an equivalent quantity** to/from the other side. The same holds true for the **multiplication or division operations**. If we multiply or divide one side of the equation by some quantity, we must also multiply or divide the other side by an **equivalent quantity**.

Concept Homework

As an assessment of your understanding of the concepts set forth in this section, answer the following questions. If necessary, review the material to help you arrive at the correct conclusions. With "true or false" statements, if the statement is true, make up an example that supports the statement without using examples given in the material. If a statement is false, correct the wording to make it true. Then make up an example that supports the statement without using examples given in the material.

1. True or False

   a. In all cases, a variable represents one and only one unknown quantity.

   b. When an equation has more than one distinct variable, the value of one variable will depend upon the values **assigned to** the other variables.

   c. Algebraic expressions will always have either addition of subtraction operations included.
      Otherwise they would be considered terms.

   d. 14 is a prime number.

   e. "Like terms" cannot have different coefficients.

   f. The expression $3x-\left(-1x+5y-4z\right)$ is equivalent to $-3x+\left[1x+\left(-5y\right)+4z\right]$

2. Explain the difference between a constant and a coefficient.

3. Explain the meaning of "like terms".

4. What is the rule for adding or subtracting like terms?

5. Explain why the subtraction operation is not commutative. Give an example.

6. What are the factors of x^3 ?

Solutions: 1a. False; 1b. True; 1c. True; 1d. False; 1e. False; 1f. False .

Exercises

1. List the **prime** factors of 12.

2. Combine like terms of each of the following:

    a. $3a + 2a - a$ b. $3ab - b + 7ab$ c. $5ax - 9bx - 11bx - 3ax$ d. $9xy - y + 8yx + 7y$

3. Change to addition and combine like terms:

    a. $2a + b - c - (3a - 2b + 3c)$ b. $4x + 7y - 3z - (-3x + 6y - 2z)$

4. Subtract the sum of $3a + 2b - c$ and $-3a - 3b - c$ from the sum of $2a - 3b - 4c$ and $5a - b - c$.

Solutions: 1. 2,2&3; 2a. $4a$; 2b. $10ab - b$; 2c. $2ax - 20bx$; 2d. $17xy + 6y$; 3a. $-a + 3b - 4c$; 3b. $7x + y - z$; 4. $7a - 3b - 3c$.

Topic 6 - Translating Words into Algebraic Expressions

The world communicates with language while mathematics uses symbols. In order for mathematics to be useful in the world, we must be able to translate words into mathematical symbols in a consistent way so that language will always be translated in the same way by people everywhere. In this topic we will become familiar with the different words and phrases that are used to express mathematical **operations** $[+, -, \cdot, \div]$.

Addition/Subtraction

There are many ways to express the concepts of addition and subtraction in language. You must become familiar with them. Here are some examples:

<u>**Words**</u>   <u>**Mathematical Translation**</u>

"added to"   as in "x added to negative 8" $x + (-8)$

"subtract"   as in   " 9 subtract 12" $9 - 12$
   Notice that in this wording, "9" comes before "12" and is translated that way mathematically!

"subtracted *from*"   as in " 9 subtracted **from** 12" $12 - 9$
   Notice that in this wording, "9" comes before "12" but <u>not so</u> in the mathematical translation!

[Note: Be careful with subtraction. The number being subtracted always <u>follows</u> the subtraction sign! As can be seen in the above subtraction examples, <u>the order</u> in which the numbers are written is *critical* to the translation. $12-9$ and $9-12$ have <u>different</u> quantitative values and are therefore <u>*not*</u> commutative.

"more than"   as in   "5 more than y" $y + 5$ or $5 + y$

"less than"   as in   "5 less than y" $y - 5$ [This is the <u>only</u> translation.]
   Notice that in this wording "5" comes before "y" but <u>not so</u> in the mathematical translation!

"plus"   as in   "p plus q" $p + q$ or $q + p$

"minus"   as in   "p minus q" $p - q$ [This is the <u>only</u> translation.]
   Notice that in this wording "p" comes before "q" <u>and also</u> in the mathematical translation!

"increased by" as in   "9 increased by y" $9 + y$ or $y + 9$

"decreased by" as in   "9 decreased by y" $9 - y$ [This is the <u>only</u> translation.]
   Notice that in this wording "9" comes before "y" <u>and also</u> in the mathematical translation!

"the sum of"   as in   "the sum of r and s" $r + s$ or $s + r$

"the difference between" as in "the difference between p and q" $p - q$ [This is the <u>only</u> translation.]
   Notice that in this wording "p" comes before "q" <u>and also</u> in the mathematical translation!

As can be seen, the addition operation is **commutative**. This is **not true** for the operation of subtraction. This becomes clear in the example $2-3$ is not equivalent to $3-2$.

Therefore, subtraction is **not** commutative. $[2-3 \Rightarrow 2+(-3) \Rightarrow -1$. **BUT** $3-2 \Rightarrow 3+(-2) \Rightarrow 1$.] That is why subtraction has **only one way** of being translated.

Multiplication

Words		Mathematics Translation
" the product of " as in "the product of 4 and a" [the result of a multiplication operation]		$4a$ [This means **4 times** *a*.]
"of" as in " $\frac{3}{5}$ of x"		$\frac{3}{5}x$ [This means $\frac{3}{5}$ **times** *x*.]
"times" as in "4 times x"		$4x$
"twice" as in " twice p "		$2p$ [This means **2 times** *p*.]

Is multiplication commutative? Yes. This can be seen easily with examples involving only numbers. $2 \cdot 3$ is equivalent to $3 \cdot 2$. This is true of **all** multiplications and is therefore **commutative**.

Division

Words		Mathematics Translation
"the quotient of" as in "the quotient of t and 16" [The result of a division operation] [t is the numerator and 16 is the denominator]		$t \div 16$ or $\frac{t}{16}$
"divided by" as in "t divided by s"		$t \div s$ or $\frac{t}{s}$
"the ratio of" as in "the ratio of 1 to 2"		$\frac{1}{2}$ [1 is the numerator and 2 is the denominator]

Notice that division is **not** commutative. $16 \div 8$ is not the same as $8 \div 16$. Also, with division in fractional form, $\frac{1}{2}$ is not the same as $\frac{2}{1}$.

Concept Homework

A. True or False If the statement is false, re-arrange it to make it true.

1. "Four less than a number" translates into mathematics as $4-n$.

2. "Four less a number" translates into mathematics as $4-n$.

3. The words "fraction", "quotient" and "ratio" all indicate division.

4. The commutative property states that the order in which quantities are inserted into an operation does not change the result of the operation.

5. The expressions $6x+10$ and $10+6x$ are equivalent.

6. The subtraction operation is commutative.   [Illustrate your answer with an example.]

7. The addition operation is commutative.   [Illustrate your answer with an example.]
8. The expressions $6x-10$ and $10-6x$ are equivalent. Show why or why not by substituting a value for x.

9. If the sum of two numbers, x and y, is 12, then y is equivalent to $12-x$. Show why or why not by substituting values for x and y.

A. Solutions: 1. False; 2. True; 3. True; 4. True; 5. True; 6. False; 7. True; 8. False; 9. True.

B. Translate the following phrases into the correct mathematical operation(s):

[After you are done, check your solutions with the answer key at the bottom of the page.]

1. Translate: **the product of twice x and 7.** _____

2. Translate: **seven greater than the product of three times y and z.** _____

3. Translate: **add 4 to the quotient of x cubed and 5.** _____

4. Translate: **the difference between x and negative 7 increased by 12.** _____

5. Translate: **the product of x subtracted from y and 3 subtracted from z.** _____

6. Translate: **7 less than twice a number, x.** _____

7. Translate: **12 decreased by the sum of x and 2.** _____

8. Translate: **The ratio of 2 and x.** _____

Solutions: $2x \cdot 7$; 2. $3yz+7$; 3. $\dfrac{x^3}{5}+4$; 4. $x-(-7)+12$; 5. $(x-y)(z-3)$; 6. $2x-7$; 7. $12-(x+2)$; 8.. $\dfrac{2}{x}$

Topic 7 - Solving Equations with a Single Variable

What do we mean by **"solving an equation"**? An equation [along with the concept of variables] is the "meat and potatoes" of algebra. Its usefulness comes from being able to model a problem from "real life" into the language of a mathematical equation where the solution lies in uncovering the value(s) of the variable(s) in the equation. Although algebra texts are full of examples and exercises in solving equations, it is important to keep in mind that the real work or algebra is in fashioning these equations from problems that occur in the world.

When we speak of "solving" and equation with a single variable ["solving for x " or whatever the variable that is being used], what we mean is:

> **"What value must x be to make the equation a true statement?"**

If we have translated a mathematical problem into an equation with a variable representing an unknown quantity [or quantities], we then look for the quantity [or quantities] that will **make the equation true**.

As was previously stated, equations with only one variable [raised to the first power] usually have a specific [only one] solution [There are two exceptions which we will discuss shortly.].

> **Example:** What number added to 3 will result in 7 ?

> Because of the simplicity of this problem, we can solve it without algebra. However, we will go through the exercise of translating the problem into an algebraic equation where the number we are looking for is defined as the variable x :

$$x + 3 = 7$$

> This is an equation with only one variable which is to the first power. It has only one solution. It is obvious that x must have the value of 4 in order to make the equation true. So, the solution is: $\qquad x = 4$

Other examples of single-variable first power equations:

$$2y - 3 = 3y + 2; \qquad \frac{3}{4}z = \frac{2}{3}; \qquad y - 2y - 3 = 4y + 2; \qquad 15 = 3p - 3.$$

As can be seen, the single variable can be represented by **any letter in the alphabet** and **might appear more than once in the equation.** Fractional coefficients and **constants** may be involved. The variable might appear **both on the left side and the right** and it **might appear more than once on either side**. It might only appear on the right side with the left side containing only a constant. **These are still examples of equations that contain only one distinct variable.** When **these types of equations** have **only one solution,** they are referred to as *conditional equations.*

Exceptions to the Rule

There are two exceptions for single variable equations to the first power having only one solution: **(1) identity equations** and **(2)equations that are a contradiction.** Neither of these types of equations are helpful in solving problems as we will see with the following examples.

Identity Equations

Equations where the **variable can take on *any* real value** and make the equation true are called **identities.**

> **Example:** $x + 2 = x + 2$

Upon inspection of this equation we see that for *any* value for x, the left side of the equation will be equivalent to the right side and the equation will be true. It is important to note, that once a value for a specific variable is fixed, it must remain consistent for that variable throughout the equation. For example, in the above equation, if we assign the value of 5 to the "x" on then left side of the equation, that same value must be assigned to the "x" on the right side of the equation.

When a variable can take on **any value** and make an equation true it is not helpful in solving a problem so the equation is **useless** in a practical sense.

Contradictions

An "equation" where there is **no value** for x that will make the equation true is called a **contradiction.**

> **Example:** $x + 2 = x + 3$

Since the x value on the left side of the equation must be consistent with the x value on the right side of the equation, we realize that no matter what value we try to assign for x, that value plus 2 cannot be equivalent to that same value plus 3. Therefore, this is not an equation in the true sense, because the left side **cannot** have the same quantitative value as the right side.

The Strategy for Solving Equations with a Single Variable

We will presently concentrate on those single variable equations [to the first power] that are neither identities nor contradictions, known as **conditional equations.** As can be seen in equation $x + 3 = 7$, $x = 4$ will make it true [The left side of the equation has the same quantitative value as the right side.]. Sometimes this solution is expressed as, "$x = 4$ **satisfies** the equation".

Any other value for x will **not** result in $x + 3 = 7$ being true. So, **"a solution for an equation"** or **"solving for x"** means: **finding the value for x that will make the equation true** [or that will "satisfy" the equation].

The basic characteristic of any equation is that the **left side has the same quantitative value as the right side,** or more simply, **the left side *always* equals the right side.**

In the example, $x + 3 = 7$, it doesn't require any manipulation to discover the solution. However, many equations are more complicated and require a strategy to discover the value of the variable. Our strategy for "solving an equation" will be to *operate* $[+,-,\times,\div]$ on the equation is such a way that we **isolate the variable** [get it alone] on **either** side of the equation. While doing this, **we must never do anything that will change the balance of the equation** [left side = right side]. Otherwise, we will have lost the value of the original equation which represents the facts from some "real life" situation that we have translated into an equation. Therefore, **we can never manipulate an equation in such a way where its balance is lost.**

To repeat, our strategy is to **isolate** the variable on **either** the left side or the right side of the equation without upsetting the equation's balance. After we have achieved this objective, we consider the **coefficient of the variable.** If the coefficient is 1, [for example $1x$] we are done. Since $1x = x$, we have found the value for x that **satisfies** the equation! But, what if the coefficient of the variable is **not** 1? [for example, $2x$, $-3x$, or $-1x$]. We must find a way to **transform the coefficient to** 1 without upsetting the balance of the equation [keeping the left side equal to the right side] .

To summarize this complete strategy:

Get the variable alone on one side of the equation with a coefficient of "1" *while still keeping the equation balanced.*

We have a number of techniques for accomplishing this task and they usually involve **inverses** and **identity elements.**

Here are some basic properties of algebra that we utilize to solve equations:

The Addition Property of Equations

Simply stated , this property is as follows:

If we add equivalent values to both sides of an equation, the equation will still be true [in balance].

[The left side will still be have the same quantitative value as the right side].

An example without variables will help illustrate the soundness of this principle:

Example 1.

Consider the following true "equation": $2+(-3)=-1$ [which is true].

When an equation has no variables, it is called a **mathematical statement.** If we add 5 to **both sides of the equation**: $2+(-3)+5=-1+5$

The "equation" [technically a mathematical statement] will **still be in balance,** or **will still be a true statement.** Adding the first two numbers of the left side results in: $-1+5=-1+5$

Completing the addition on both left and right side: $4=4$

which , shows that **the "equation" is still true!**

Would the statement still be true if we added the opposite of 5 to both sides? Try adding -5 to both sides of $2+-3=-1$. Would the mathematical statement still be in balance?

The answer is "yes". In fact, the more examples we try with different values being added, the more we realize that we can add **any** equal value to both sides and it will **remain a true statement**.

It is important to remember that this property allows us to add **any** equal value to both sides of an equation [equations contain variables]. As we will see in the following examples, we get to **choose** a **_convenient_** quantity that will help us accomplish our objective of discovering the value of a variable.

Now let's look at an equation with a variable:

Example 2.

Solve for x: $x-5=-9$

The first thing we do is put in the "implied" coefficient of x, making it $1x$. [Multiplying by 1, the identity element for multiplication, does not change the value of x.] Next, we change subtraction to addition: $1x+(-5)=-9$

We now utilize **the Addition Property of Equations** by adding 5

[the additive inverse or opposite of -5] to **both** sides of the equation:

$$1x+(-5)=-9$$
$$\underline{\qquad 5 \qquad 5}$$
$$1x + 0 = -4$$

We **_arbitrarily chose_** 5 because it's the **inverse** [opposite] of -5. When we add them, we get 0, the **identity element for addition**. This, in effect, eliminates -5 from the left side of the equation and **isolates** x [gets it alone]. Because it was added to **both** sides of the equation, the equation is still **true** [or "balanced"].

Adding 0, the identity element for addition, doesn't change the
value of $1x$, so we can eliminate 0: $\qquad\qquad 1x = -4$
Or just: $\qquad\qquad x = -4$
[solution]

Note: $x = -4$ is the **solution** to the equation because substituting -4 for x in the original form of the equation, will **make it true** or **balanced** or **makes the left side is equivalent to right side** or **satisfies** it. It is the **only** value for x that will make this equation true. Still another way of expressing this idea is to say that $x = -4$ **solves the equation,** $x - 5 = -9$.

We can also use this *Addition Property of Equations* with variables as well as constants.

Example 3.
Solve for x: $\qquad 4x + 6 = 3x$

We can eliminate $3x$ from the right side of the equation by adding $-3x$ to **both sides** using the **Addition Property of Equations**. It

pertains to **variables** as well as **constants**:

$$4x + 6 = 3x$$
$$\underline{-3x \qquad\quad -3x}$$

On the right side of the equation, $3x$ and $-3x$, are **opposites** [additive inverses]. **Any value** that we assign to x will make $3x$ and $-3x$ have opposite values which means their addition results in 0. Combining like terms results in: $\qquad 1x + 6 = 0$

Now the variable term is only on one side of the equation. Having 0 on the other side of the equation is mathematically valid [For example $5 + (-5) = 0$]. It is still a balanced equation even though one side has the value of 0. [**Having the value of 0 is not the same as having "no" value. 0 only "disappears" when it is being added to another value.**] We now add -6 to both sides of the equation [**addition property of equations**] which eliminates the constant, 6 from the **left side** of the equation:

$$1x + 6 = 0$$
$$\underline{-6 \quad -6}$$

The 6 and -6 on the left side are opposites [**additive inverses**] which make them equivalent to 0: $\qquad 1x + 0 = -6$

The 0 "disappears" as it is the identity element for addition: $\qquad 1x = -6$
We can omit the coefficient, "1" as it is the
identity element for multiplication: $\qquad x = -6$
[solution]

Identity equations

Earlier in this topic we talked about identity equations where the variable can take on **any** value and the equation will be true. Here is an example of what happens if we try to solve such an equation.

Example 4.

Solve for x:

$$3x - 2x + 5 = 4x - 3x + 4 + 1$$

First we will **combine like terms** [underlined above] on each side of the equation:

$$1x + 5 = 1x + 5$$

Now, if we *choose* to add $-1x$ to both sides of the equation:

$$-1x \qquad -1x$$

The result is:

$$0 + 5 = 0 \qquad +5$$

Or just:

$$5 = 5$$

This is a true mathematical statement **but the variable, x, has been eliminated from the equation**. This tells us that x could have been *any value* and the **equation would be mathematical true**. Let's go back to the first step that we made:

$$1x + 5 = 1x + 5$$

Suppose we chose to add the opposite of 5 to each side of the equation:

$$-5 \qquad -5$$

The result would be:

$$1x + 0 = 1x \qquad +0$$

Adding 0, the identity element for addition doesn't affect value, so:

$$1x = 1x$$

Or simply:

$$x = x$$

Again, this is an equation that will be true regardless of the value of x.

When **any value** can be substituted for the variable in an equation, the equation is valueless. Just imagine a situation where the value we are trying to uncover could be anything!

Equations that are contradictions

As we explained earlier, these equations cannot be solved since there are **no values** for the variable that will make the equation true. Here is an example of what happens if we try to solve such an equation.

Example 5.   **Solve for** x: $\underline{3x-2x}+5=\underline{4x-3x}+4$

First we will combine like terms **[underlined above]** on
each side of the equation:

$$1x\ +5=\ 1x\ +4$$

Now, if we add $-1x$ to both sides of the equation:

$$\underline{-1x \qquad\qquad -1x}$$

The result would be:

$$0\ +5=\ 0\ +4$$

Adding 0, the identity element for addition doesn't affect
value, so:

$$5=4$$

This, of course, is **not a true mathematical statement** and
and will never be true regardless of the value
that x assumes. Let's go back to the first step that we made:

$$1x\ +5=\ 1x\ +4$$

Suppose we chose to remove 5 from the left side of the
equation by adding -5 to both sides of the equation:

$$\underline{-5 \qquad\qquad\qquad -5}$$

The result would be: $1x\ +0=1x\ +\left(-1\right)$

Or more simply: $x\ =x\ -1$

We can easily see that there is no value that we can give for x on both sides of the equation
[it would have to be the same for both x's **]** that would make this a true statement. Therefore the
equation **cannot be solved** and is a **contradiction**.

We will now continue with examples of **conditional equations** that give use **useful** results.

The Multiplication/Division Property of Equations

As you might suspect, this idea has similarities to the addition property:

If we multiply or divide *every* **term on** *both* **sides of an equation by the same quantity,
the equation will still be true [in balance].**

[The left side will still be equivalent to the right side, another way of saying that the equation will still be "true".]

A simple example without variables will help illustrate this principle:

Example 6.   Consider the following mathematical statement: $2+\left(-3\right)=-1$

If we multiply **every term in the equation** by -3, the
equation will ***still be in balance*** , or will ***still be true***: $\underline{-3}\cdot 2\ +\ \underline{\left(-3\right)}\left(-3\right)\ =\ \underline{-3}\left(-1\right)$

Simplifying both sides of the equation results in: $-6\ +\ 9\ =\ 3$

Simplifying: $3\ =\ 3$

[The "equation" is still balanced.]

60

We can also use *division* in the same manner:

Consider the following mathematical statement ["equation" without variables]:

$$8 + (-2) = 6$$

If we divide ***every term*** in the equation by -2, the equation will still be in balance, or will still be true:

$$\frac{8}{-2} + \frac{-2}{-2} = \frac{6}{-2}$$

Performing the division results in: $-4 + 1 = -3$

Then performing the addition: $-3 = -3$

[the "equation" is still balanced.]

We could have used **any** divisor in these examples and would have ended up with a true mathematical statement!

Often, we use **both** the **Addition Property of Equations** and the **Multiplication/Division Property of Equations** to isolate the variable and give it a coefficient of 1 [solving the equation.]:

Example 7.   Solve for x: $\frac{3}{5}x + 3 = 6$

We start by dealing with the constant, 3. We utilize the **addition property of equations** by adding -3 [the inverse of "opposite" of 3] to

both sides of the equation:

$$\frac{3}{5}x + 3 = 6$$
$$\underline{\quad\quad -3 \quad -3\quad}$$

This results in: $\frac{3}{5}x + 0 = 3$

"0" is the identity element for addition, so it can be eliminated: $\frac{3}{5}x = 3$

Now we have x alone on one side of the equation. However, its **coefficient is not** 1 [which it needs to be!]. We can accomplish this by multiplying **every term on both sides of the equation** by $\frac{5}{3}$, the

reciprocal [inverse for multiplication] of $\frac{3}{5}$: $\frac{5}{3} \cdot \frac{3}{5}x = \frac{5}{3} \cdot 3$

Since the reciprocal of $\frac{3}{5}$ is its **inverse**, the result of this

multiplication is the **identity element for multiplication.** $\left(\frac{5}{3} \cdot \frac{3}{5} = \frac{15}{15} \text{ or } 1\right)$.

{ 60 }

We have accomplished the goal of **making**

$$\text{the coefficient of } x \text{ equal to } 1: \qquad \frac{15}{15}x = \frac{5}{3} \cdot 3$$

$$\text{Which is the same as:} \qquad 1x = \frac{5}{3} \cdot 3$$

Now, let's simplify the right side of the equation.

$$\text{We'll start by transforming } 3 \text{ into its equivalent fraction, } \frac{3}{1}: \qquad 1x = \frac{5}{\cancel{3}} \cdot \frac{\cancel{3}}{1}$$

$$\text{Eliminating the identity element, } \left[\frac{3}{3}\right], \text{ or "canceling the 3's":} \qquad 1x = \frac{5}{1}$$

$$\text{Changing the fraction to its equivalent whole number:} \qquad 1x = 5$$

$$\text{Or in its more conventional form:} \qquad x = 5$$

[solution]

Therefore, $x = 5$, **satisfies** the equation, $\frac{3}{5}x + 3 = 6$ **[makes the equation true]**.

Using the Distributive Property in Solving Equations

Sometimes the distributive property must also be utilized in solving equations. In general, this property is symbolized by $a(b+c) = ab + ac$. Notice that both b and c are multiplied by a and the results are added. The variables, $a, b \, \& \, c$, represent any real numbers.

Example 8.   **Solve for** x: $5(3x-1) + 2 = 12x + 6$

Now, the **distributive property** is employed to simplify the left side of the equation. In this case 5 must multiply both $3x$ and 1 **[the numbers inside the parenthesis]**: $\qquad \underline{\underline{5 \cdot 3x}} - \underline{\underline{5 \cdot 1}} + 2 = 12x + 6$

$$\text{The multiplication results in:} \qquad 15x - 5 + 2 = 12x + 6$$

$$\text{We now change subtraction to addition on the left side:} \qquad 15x + \underline{\underline{(-5)}} + 2 = 12x + 6$$

The ultimate goal is to isolate the x term on either side of the equation. The first step is to further simplify the equation by adding the constants **[underlined above]** on the left side of the equation: $\qquad 15x + (-3) = 12x + 6$

Using the addition property, we add 3 to both sides of the equation, to eliminate the constant, -3, on the left side of the equation: $\qquad \dfrac{3 \qquad\qquad 3}{}$

$$\text{Resulting in:} \qquad 15x \qquad = 12x + 9$$

$$15x \quad = 12x + 9$$

Now we can remove the $12x$ term from the right side of the equation by adding its opposite to both sides of the equation:

$$-12x \qquad -12x$$

[Notice that we can use the addition property with terms that contain variables as well as constants!]

Resulting in:

$$3x \quad = \quad 9$$

Finally, we divide both sides of the equation, $3x = 9$, by 3 to transform the coefficient of the x term to 1:

$$\frac{3x}{3} = \frac{9}{3}$$

This results in:

$$1x = 3$$

Or more simply:

$$x = 3$$

The solution, $x = 3$, tells us that when 3 is substituted into the original equation, $5(3x - 1) + 2 = 12x + 6$, the equation will be true.

Removing Fractions from an Equation

Many equations will have fractional coefficients within them. We have already seen the technique of multiplying the equation by the fraction's reciprocal to eliminate the fraction. When there is **more than one** fractional coefficient or term, another technique involving the **multiplication property** is employed to **clear all the fractions in one step**. It is important to remember that *every* term on both sides of the equation must be multiplied when using this property.

Example 9.   **Solve for x:** $\frac{1}{2}x - 1 = \frac{2}{3}x - 4$

If we multiply this equation by a number that both denominators will divide evenly, we will be able to **eliminate both denominators** through the "canceling" process. The smallest such number in this case is 6 **[6 is the smallest number that both the denominators, 2 and 3 will divide evenly]**. Therefore we will use 6 as the multiplier:

$$6\left(\frac{1}{2}x - 1\right) = 6\left(\frac{2}{3}x - 4\right)$$

Using the distributive property:

$$\frac{6}{1} \cdot \frac{1}{2}x - 6 \cdot 1 = \frac{6}{1} \cdot \frac{2}{3}x - 6 \cdot 4$$

Notice that we substituted $\frac{6}{1}$ for 6 when multiplying the fractions. Since $\frac{6}{1}$ and 6 have the same value they are **interchangeable**. By doing this, fractions are multiplying fractions and whole numbers are multiplying whole numbers. Canceling [removing identity elements]:

$$\frac{\overset{3}{\cancel{6}}}{1}\cdot\frac{1}{\underset{1}{\cancel{2}}}x-6\cdot1=\frac{\overset{2}{\cancel{6}}}{1}\cdot\frac{2}{\underset{1}{\cancel{3}}}x-6\cdot4$$

Results in: $3\cdot1x-6\cdot1=2\cdot2x-6\cdot4$

Now the fractions have been cleared since fractions with denominators of 1 are interchangeable with their integer equivalents. This greatly simplifies the equation.

Performing the remaining multiplications results in: $3x-6=4x-24$

We now **isolate the variable**. First we will add 6 to both sides of the equation: $\qquad \underline{6 \qquad\qquad 6}$

This removes the -6 from the left side of the equation: $3x \qquad = 4x-18$

We now get the x terms all on the left side of the equation by adding $-4x$ to both sides: $\underline{-4x \qquad -4x}$

Resulting in: $-1x \quad = \quad -18$

We need to transform the $-1x$ to $1x$. We can do this by either multiplying or dividing both sides of the equation by -1. Either strategy will accomplish the goal. In this case, we will multiply both sides of the equation by -1: $\quad (-1)(-1x)=(-1)(-18)$

Performing the multiplication results in: $1x=18$

Or simply: $x=18$

[solution]

This means that when we substitute 18 for x in the original equation, $\frac{1}{2}x-1=\frac{2}{3}x-4$, the equation will be true.

Let's now consider how we can solve a problem given in words by translating them into an **algebraic equation.**

Example 10.

One number is 4 less than twice another number. Their sum is negative twenty-two. What are the numbers?

Step 1

The first question we **always** ask ourselves in an "application problem" is "**What are we trying to find?**"

In this case, we are trying to find the value of **two numbers that are** *related to each other*.

Step 2

Our first instincts would be to think that we need two variables since we are trying to find two different numbers. However, since the two numbers are *related*, **we only need one variable !** The problem tells us that "one number is 4 less than twice another number". Therefore. if we find "another number" we need only to multiply it by 2 and subtract 4 to find the "one number".

Now we need to assign a variable [**we will use** x] to represent **"another number"**.

We say: Let $x=$ "another number"

Now, **in terms of** x, and using "4 less than twice **another number**", which we have assigned to x , "**one number**" would be $2x-4$. So, we also say: Let $2x-4=$ "one number"

Step 3

We now want to **create an equation**. The problem tells us that both numbers together sum to -22. Therefore, the equation would be : $x+2x-4=-22$

When we solve for x, we will know the "another number" . We can start be giving x its "implied" coefficient [1] and change subtraction to addition: $1x+2x+(-4)=-22$

Next we add the **like terms** [on the left side]. Having "$1x$" instead of "x" facilitates this process: $3x+(-4)=-22$

Our objective is to **isolate the variable**. The **constant** $[-4]$ needs to be eliminated from the left side of the equation where the x term resides. We accomplish this with the **addition property of equations.** We add

4 to **both sides** of the equation:

$$3x+(-4)=-22$$
$$\underline{44}$$
Resulting in: $3x + 0 = -18$

The identity element for addition, 0, can be eliminated, so what remains is: $3x = -18$

We now must use **the division property of equations**.

We divide *every* term in the equation by 3:
$$\frac{3x}{3} = \frac{-18}{3}$$

This creates the identity element **[for multiplication]**, $\frac{3}{3}$, as

the coefficient of x which we change to its equivalent, 1:
$$1x = -6$$

Since $1x$ and x are **interchangeable** we can just say:
$$x = -6$$
[one solution]

So, we have found that the **"another number"** is -6.
Substituting that value into $2x - 4$ will give us the
"one number". Let $x = -6$ in:
$$2x - 4$$

Results in:
$$2(-6) - 4$$
Doing the multiplication first:
$$-12 - 4$$
Changing subtraction to addition:
$$-12 + (-4)$$
Adding results in the solution to the **"one number"**:
$$-16$$
[the other solution]

Therefore the two numbers we are looking for are -6 and -16. We can see that this makes sense since $-6 + (-16) = -22$ which **satisfies** the original equation **[makes it true]**.

Concept Homework

As an assessment of your understanding of the concepts set forth in this section, answer the following questions. If necessary, review the material to help you arrive at the correct conclusions. With "true or false" statements, if the statement is true, make up an example that supports the statement without using examples given in the material. If a statement is false, correct the wording to make it true. Then make up an example that supports the statement without using examples given in the material.

1. True of False:
   a. We eliminate a constant from one side of an equation by adding its additive inverse [its *opposite*] to both sides of the equation.

   b. In order to utilize the **Multiplication/Division Property of Equations**, we must multiply or divide **every term** on both sides of the equation by the same quantity.

   c. We often employ **Division Property of Equations** to change a coefficient of a variable to "1".

   Solutions: 1a. True; 1b. True; 1c. True.

2. When we find the value of the variable(s) in an equation, it is said that we have solved the equation. Explain what that means to have solved an equation.

3. Give an original example of an equation that has an infinite number of solutions.

4. Give an original example of an equation that has no solution.

5. Explain the idea behind the **Addition Property of Equations**.

6. How are additive inverses [opposites] used in solving equations? Then, give an original example.

7. State the **Multiplication/Division Property of Equations.**

8. What is another name for a multiplicative inverse [inverse for multiplication]? Explain how are they used in solving equations? Then give an original example.

9. What is the **distributive property of mathematics**? Give an example of its use.

10. When we have multiple fractional coefficients in an equation, how can we eliminate them in one step?

Exercises

Find the solutions to the following equations. If there are an infinite number of solutions, state that the equation is an **identity**. If there are no solutions, state that the equation is a **contradiction**.

1. $2x + 7 = 11$　　**2.** $7x - 3 = -17$　　**3.** $x - 2x + 12 = 12 - x$　**4.** $5x = 8 + 3x$

5. $6 + 5x - 2 = 4 - 5x$　**6.** $\dfrac{3x}{2} + \dfrac{1}{6} = \dfrac{2x}{3} - \dfrac{2}{3}$　**7.** $x + 4 - 2x = x + 3 - 2x + 2$

8. $\dfrac{x}{4} - \dfrac{x}{12} = \dfrac{x}{2} + \dfrac{1}{2}$　　**9.** $5 + 3(2x + 1) = 0$　　**10.** $3(7x - 2) = 11 - 4(2x - 3)$

11. When 15 is added to a number, the result is 21. Find the number.

12. One number is one less than 7 times another number. Their sum is 31. Find the two numbers.

13. The difference between one-third of an integer and one-fourth of the same integer is 3. Find the integer.

Solutions: 1. $x = 2$; 2. $x = -2$; 3. identity equation; 4. $x = 4$; 5. $x = 0$; 6. $x = -1$; 7. contradiction; 8. $x = -\dfrac{3}{2}$; 9. $x = -\dfrac{4}{3}$; 10. $x = 1$; 11. The number is 6 ; 12. One number is 4 and the another number is 27; 13. The integer is 36.

Topic 8 - Applications for Equations with One Variable

As previously mentioned, for algebra to be useful, we must be able to use it solve problems occurring in the "real world". The strategy is to convert the facts given in a real life situation into an equation, usually with assigning a variable to represent the number which we are trying to determine.

The following are some examples of such problems where the strategy is laid out and executed.

Example 1.

A plumber charges $26.00 per hour. He spends 30 hours plumbing a new bathroom for a customer's home. The total cost of labor and materials for the job was $1,390. How much did the materials cost?

Although this problem could be solved without algebra, it is good practice to structure a solution by defining a variable and creating an equation to arrive at a solution.

What is our basic strategy for accomplishing this?

- First, it is always a good idea to read the problem a few times so we become familiar with the facts given and the question to be answered.

- **What is the question to be answered?** This is usually what will define the variable. We wish to know the cost of material, so let m =**cost of material**.

- The next step is to **create and equation**. We know that the entire cost is $1390. That will become the **right side of the equation.** The **left side of the equation** is labor cost plus material cost. We do not need a variable to represent the labor cost since that can be determined by the facts given in the equation $[30 \text{ hrs.} \cdot \$26 \text{ per hour}]$. The material cost [what we are trying to find] is the variable, m.

We are now ready to **write the equation:**	$30 \cdot 26 + m = 1,390$
Performing the multiplication:	$780 + m = 1,390$
Using the addition property of equations, we add -780 to both sides of the equation:	$-780 \qquad -780$
resulting in :	$m = 610$

Solution: The cost of materials is $610.

As previously stated, we could have arrived at the solution through a logical process that is not algebraic. **However, algebra forces us to organize our thinking in a structured way.**

Example 2.

Here is an example of a "mixture problem" that is most easily solved using an algebraic approach:

> **How many pounds of sunflower seeds that cost $1.75 per pound must be mixed with 30 pounds of cashews that cost $5.00 per pound to make a mixture that costs $3.00 per pound?**

First, we determine what we are asked to find [How many pounds of sunflower seeds]. This defines our variable. **Let $x =$ the number of pounds of sunflower seeds.**

Now we must **create an equation** with the given information. The equation must include the **cost** [dollar amount] of each kind of ingredient [number of lbs. times the price per pound]. On the **left side of the equation** we have x pounds of sunflower seeds at $1.75 per pound along with 30 pounds of cashews at $5.00 per pound. Therefore, the **left side of the equation** is:

$$1.75 \cdot x + 30 \cdot 5.00$$

Now we need to create an **equivalent amount** for the right side of the equation. We know that this side of the equation must cost $3.00 per pound. The question then becomes, "How many pounds must there be on the right side of the equation?" There are x pounds of sunflower seeds and 30 pounds of cashews. This totals $(x + 30)$ at $3.00 per pound. Therefore, an **equivalent amount for the right side** of the equation is:

$$3.00(x + 30)$$

Now we need to solve this equation that we have created to determine the value of x [the amount of sunflower seeds needed]:

Solve for x:

$$1.75x + 5.00 \cdot 30 = 3.00(x + 30)$$

We start by simplifying the left side and distributing the right side:

$$1.75x + 150 = 3.00x + 90$$

We now must isolate the x terms on one side of the equation and the constants on the other.

Upon inspection, we see that keeping the x terms on the right side and the constants on the left side, will eliminate the need for having a negative x term. Although we are used to seeing the x terms on the left side of an equation, it is not necessary for the equation to remain valid. Therefore

$$1.75x + 150 = 3.00x + 90$$

we will add $-1.75x$ to both sides of the equation:

[This is the addition property of equations].

$$\underline{-1.75x \qquad\qquad -1.75x}$$

This results in:

$$150 = 1.25x + 90$$

We now add -90 to both sides of the equation so that the x term will be **isolated**:

$$\underline{-90 \qquad\qquad -90}$$

Resulting in:

$$60 = 1.25x$$

We have to transform the coefficient of x to 1 without disrupting the balance of the equation. We accomplish this by dividing both sides of the equation by 1.25:

$$\frac{60}{1.25} = \frac{1.25x}{1.25}$$

[This is the division property of equations].

This creates the coefficient of 1 for the x term and gives us the solution to the problem:

$$48 = 1x$$

or more conventionally:

$$x = 48$$

Going back to the definition of x, and the question posed in the problem:

We need 48 pounds of sunflower seeds for a mixture that will cost \$3.00 per pound.

Example 3.

It will be good practice to look at the same problem from a different perspective:

> **A nut distributer wishes to create a mix of sunflower seeds and cashew nuts totaling 78 pounds that will cost \$3.00 per pound. If sunflower seeds cost \$1.75 per pound and cashew nuts cost \$5.00 per pound, how many pounds of each should be in the mix?**

First we determine what is being asked. We need to know two things: the number of lbs. of sunflower seeds and the number of lbs. of cashew nuts. Our first instinct is to think that we need two variables. However, since we know the total number of pounds [78], **there is a relationship** between the number of lbs. of sunflower seeds and cashews. Therefore, we can solve the problem with only one variable!

Let x = the number of pounds of sunflower seeds.

Since the total pounds are 78, the number of pounds of cashews becomes 78 less the number of pounds of sunflower seeds. Therefore:

Let $78 - x$ = the number of pounds of cashew nuts.

We can now create an equations using both the number of lbs. of each ingredient along with the cost of each. **The right side of the equation** is spelled out for us in the problem. There is to be 78 pounds at \$3.00 per pound. $\left[3.00 \cdot 78\right]$

The left side of the equation has to be equivalent to this but is broken up into sunflower seeds and cashew nuts. Using the variable defined above and the costs per pound of each ingredient given to us in the problem, we arrive at $1.75x + 5.00(78 - x)$.

Therefore, the equation becomes:	$1.75x + 5.00(78 - x) = 3 \cdot 78$
Distributing the left side of the equation:	$1.75x + 5.00 \cdot 78 - 5.00 \cdot x = 3 \cdot 78$
Performing the multiplication:	$1.75x + 390 - 5.00x = 234$
Combining like terms [$1.75x - 5.00x$] on the left side:	$-3.25x + 390 = 234$
Adding -390 to both sides of the equation:	$-390 \quad -390$
Results in:	$-3.25x \qquad = -156$

We now need to transform the coefficient of the x term to 1. We accomplish this by dividing both sides of the equation by -3.25:

$$\frac{-3.25x}{-3.25} = \frac{-156}{-3.25}$$

Resulting in: $\quad 1x \;=\; 48$

or just: $\quad x = 48$

From our definition of x, we conclude that **we need 48 pounds of sunflower seeds.**

We also need to determine the amount of cashews which we defined as $78 - x$. Substituting 48 for x, we arrive at $78 - 48 = 30$.

Therefore, **we need 30 pounds of cashews.**

Example 4.

A teenager robbed his piggy bank so that he could take his new girlfriend out to lunch at a fast food restaurant. His bill came to $7.75. He gave the cashier 55 coins. All were either nickels or quarters. How many nickels and quarters did the cashier receive?

At first glance, it would seem that we would need two variables to solve this problem [one for nickels and one for quarters]. However, since we know the total number of coins, **there is a relationship** between the number of nickels and the number of quarters. Therefore, we only need one variable.

Defining the variable: Let us say that n represents the number of nickels. Then, if there are 55 coins all together, the number of quarters must be 55 less the number of nickels! Therefore, $\quad n =$ **the number of nickels** and

$\qquad 55 - n =$ **the number of quarters**

Creating an equation: We are given the total monetary value of the 55 coins as 7.55. We will use this information to create an equation. The **right side of the equation** will therefore be 7.55. How do we make the left side of the equation equivalent using the variable that we have created? The monetary value of a nickel is 0.05 [5 cents] while a quarter is 0.25 [25 cents]. If we multiply the number of nickels and quarters by their monetary value, they will be equivalent to 7.55. The left side of the equation becomes $.05x + .25(55 - x)$

We are now ready to create the equation and solve it for x.

Solve for x:	$.05x + .25(55 - x) = 7.55$
We start by distributing on the left side:	$.05x + .25 \cdot 55 - .25 \cdot x = 7.55$
Doing the multiplication:	$.05x + 13.75 - .25x = 7.55$

At this point, it can become less confusing if we were to eliminate the decimals. We can accomplish this be multiplying the **entire** equation by 100.

This does **not** change the **balance** of the equation! $5x + 1375 - 25x = 755$
[Notice that multiplying each term by 100 moved the decimals two places to the right!]

Combining the "like" x terms on the left side: $\quad -20x + 1375 = 755$

We then isolate the x term by adding -1375 to both

sides of the equation: $\qquad \underline{-1375 \quad -1375}$

Results in: $\qquad -20x \qquad = -620$

In order to transform the coefficient of x to 1, we

$$\frac{-20x}{-20} = \frac{-620}{-20}$$

divide both sides of the equation by -20:

Resulting in: $1x = 31$

Or more conventionally: $x = 31$

So the number of nickels [represented by x] **is 31**

The number of quarters [represented by $55 - x$] **is** $55 - 31$ **or** 24

Therefore, the cashier received 31 **nickels and** 24 **quarters**

Notice that if we add the total number of coins it sums to 55, which corresponds with the facts given in the problem. Also, 31 nickels has the value of $1.55 and 24 quarters has the value of $6.00. If we add these values together, we get $7.55 which is what it needs to be. It would be possible to try different combinations of 55 nickels and quarters until we hit upon the right amount of money [$7.55] . As can be seen this trial and error process could take a long time, and if the numbers in the problem were larger, it would take even longer. Using algebra is a much more efficient approach.

Algebra is a useful problem solving tool. The most creative part of the process [and most difficult] is applying the process of **defining variables** and **creating an equation** to find a solution(s) for "real world" problems.

Example 5.

This next problem is an example of what are generally categorized as **"motion problems"**. They are based on the formula:

$$\textbf{D}\text{istance} = \textbf{R}\text{ate} \cdot \textbf{T}\text{ime}$$

Using this formula, when we want to calculate the distance traveled, we would multiply the rate of speed [miles/km per hour] by the number of hours traveled. For example, if we drive at an average rate of speed of 60 MPH for two hours, we would have traveled 120 miles [$60 \cdot 2$] .

We can also use this basic formula in solving more complex "motion problems" as shown in this example:

Two cars on the same road are 464 miles apart and are traveling towards each other. One is traveling 8 miles per hour [MPH] faster than the other. If they will meet in 4 hours, how fast is each car traveling?

We start by reading the problem over a few times until we have sorted out all the information given in the problem. We then ask ourselves, '**What are we trying to find?**'. The answer will steer us in the right direction for defining the variable that will help solve the problem. We need to find the rates [speeds] of each car.

Let $x =$ the rate [MPH] of **one** of the cars.

We are told that one car is traveling 8 MPH faster than the other. Therefore, we can define the speed of the other car as:

Let $x + 8 =$ the rate [MPH] of the **other** car.

Since the cars are traveling towards each other, their combined rate in is found by **adding their speeds together**:

Total rate $= x + x + 8$ or $2x + 8$

We can now use our basic motion formula to solve the problem:

Distance = **R**ate · **T**ime

Using the variable definitions made above along with information given to us in the problem we can substitute values for **D**istance, **R**ate and **T**ime:

Distance: 464 miles **R**ate: $2x + 8$ (MPH) **T**ime: 4 hours

We can now write an equation with the only variable, x, which, give us the answer to the questions asked in the problem:

$$\text{Distance} = \text{Rate} \cdot \text{Time}$$
Substituting: $\quad 464 = (2x + 8) \cdot 4$

We use the **commutative property** on the right side of the equation to make it more conventional prior to using the distributive property: $\quad 464 = 4(2x + 8)$

Performing the distribution on the right side: $\quad 464 = 8x + 32$

Using the addition property, we add -32 to both sides:

$$464 = 8x + 32$$
$$\underline{-32 \qquad -32}$$

Resulting in: $\qquad 432 = 8x$

Using the division property, we divide both sides of

the equation by 8: $\qquad \dfrac{432}{8} = \dfrac{8x}{8}$

Results in: $\qquad 54 = x$

Or more conventionally: $\qquad x = 54$

x was defined as the speed of **one** car (54 MPH).
The **other** car's speed was defined as $x + 8$: (54+8 MPH) or 62 MPH.

So, **one** car was traveling at 54 MPH while the **other** was traveling at 62 MPH [solution]

Strategies for Solving Problems Using Algebraic Techniques

- **Start by reading the problem several times so that the information given as well as the problem to be solved is clear.**

- **The variable will usually be defined by what question we are asked to answer. If there is more than one question to be answered, we try to find a relationship between the questions so that they can all be defined with only one variable.**

- **By translating the information given and the defined variables, we create an equation whose solution will solve the problem.**

Exercises

1. In a magical performance, Alphonse the Magnificent is given a box containing only dimes and quarters. He is informed that there are 24 coins in the box that have a total value of $4.50. He makes some calculations and comes up with the correct number of coins and claims that he has used his knowledge of algebra to solving this problem. What was his solution and how did he arrive at it?

2. A theatre sold 800 tickets for a total of $2,275.00. Some tickets cost $2.50 and some cost $3.50. How many of each kind were sold?

3. A grocer mixes two kinds of coffee beans, one worth $4.75 per pound and the other worth $3.25 per pound. If the mixture weighs 300 pounds and sells for $3.95 per pound, how many pounds of each kind does he use?

4. How many pounds of tea worth $7.00 per pound must be mixed with 27 pounds of tea worth $5.75 per pound to make a mixture that sells for $6.25 per pound?

5. Two cars starting at the same place travel in opposite directions. The first car averages 45 MPH while the second averages 50MPH. In how many hours will the cars be 570 miles apart?

Solutions: 1 . 14 quarters and 10 dimes; 2. 525 tickets @ $2.50 & 275@$3.50; 3. 140 lbs.@$4.75 and 160 lbs.@$3.25; 4. 18 lbs.; 5. 6 hours.

Topic 9 - Solving Literal Equations

Some equations have more than one solution. For example, if we want to find the area of a rectangle, we need to know its length and the width. Every rectangle can be sized differently, so the area can take on many different values depending on its length and width. What all these solutions have in common is that their areas all can be determined by multiplying the length times the width. Expressed as an equation in terms of variables, the equation would be $A = lw$ [Since there is no indication of an operation between l and w, it is assumed to be multiplication.]

This equation differs from the single variable equations that we discussed in the previous section as it has three distinct variables [A, l and w]. When equations have more than one distinct variable, they fall into the category of **literal equations**. Finding the area of a rectangle [a literal equation] requires knowing the length and the width. For *any* literal equation, finding the value one of the variables requires that we **assign values** to all the other variables. As opposed to single variable equations, each variable in a literal equation can take on *many* values **depending upon the values assigned to the other variables**.

Another name for a literal equation is a **formula**. $A = lw$, as well as thousands of other literal equations or formulas, determines the **relationship between variables.**

Example 1.

Consider the literal equation: $c = 2a + b$

The value of c **depends** on the values assigned to a and b. If $a = -2$ and $b = 5$ then we can determine the value of c for this case by

substituting these values into the equation: $c = 2(-2) + 5$

Performing the multiplication first [as required by "order of operations"]: $c = -4 + 5$

The addition results in: $c = 1$

[Solution]

What we are saying is:

When $a = -2$, $b = 5$, and $c = 1$, the equation, $c = 2a + b$, **is true.**

Every time we change the value of one of these variables, the other variables will **also change.** Therefore, we cannot determine a unique [only one] value of any one of these variables unless we know the values of the other two. Suppose the values of **all** the variables in this literal equation remain unknown. Then all we know is that a, b and c are **related** in such a way that $c = 2a + b$. If we wanted to solve the equation for either a or b, we would **isolate** that variable on one side of the equation.

Example 2.

Suppose, with the equation, $c = 2a + b$, we wanted to get b alone on one side of the equation. We call this **"solving the equation for b in terms of a and c"**. This is the only way of "solving" a literal equation when the values of the variables are unknown.

We would accomplish this by **manipulating** the equation just as we do equations in only one variable. The goal is to have b alone on one side of the equation with a coefficient of **"1"** [which it already is in this case]. This is done very much the same way as if b were the **only variable in the equation**.

We start with: $\qquad c \quad = \quad 2a + b$

Using the **addition property of equations**,
we add $-2a$ to **both** sides of the equation: $\qquad \underline{-2a \qquad -2a}$

Notice that we cannot combine $2a$ and c on the left side of the equation because they are not like terms. However, $2a$ and $-2a$ on the right side of the equation **are** like terms. In fact, they are **opposites.** So they add to 0 on the right side of the equation. This results in: $\qquad c - 2a = 0 + b$

As "0", the **identity element for addition**, does not change the value of b, it is no longer need it in the equation: $\qquad c - 2a = b$

b **is now by itself on the** right side **of the equation!**

We now interchange the left and right sides or the equation. We can do this because in **any** equation, the left side **is equivalent to the right side** and, therefore, **interchangeable:** $\qquad b = -2a + c$

[Solution]

We have "solved" the equation for b *in terms of a and c* !

Example 3.

Suppose we wanted to solve the same equation, $c = 2a + b$, for a **in terms of** b **and** c. This merely means that we wish to **isolate** a **on one side of the equation.**

We start with: $\qquad c = 2a + \ b$

Because we are solving for a, we want to **isolate** it [get it alone on one side of the equation]. Therefore, we utilize the **addition property of equations** to remove b from the right side by adding $-b$ to both sides: $\qquad \dfrac{-b \qquad\qquad -b}{-b + c = 2a}$

results in:

We now **interchange** the **equivalent** left and right sides: $\qquad 2a = -b + c$

However, we are not finished, because we need $1a$ instead of $2a$ on the left side [The coefficient of a has to be 1 to "solve" the equation.].

We accomplish this by dividing **every** term in the equation by 2: $\qquad \dfrac{2a}{2} = \dfrac{-1b}{2} + \dfrac{1c}{2}$

[Note: We have added the implied coefficient "1" to the b and c variables for clarity.]

This creates an identity element $\left[\dfrac{2}{2}\right]$ as the coefficient of a which

is equal to 1: $\qquad 1a = \dfrac{-1b}{2} + \dfrac{1c}{2}$

Or its equivalent: $\qquad a = \dfrac{-b + c}{2}$

[solution]

Note: On the right side of the equation, the denominators are the same, so we can combine the fractions by adding the numerators. If necessary, review the rules for adding fractions. They are the same whether they have variables or not!

Example 4. - Application

As we previously emphasized, to make algebra useful, we must translate "real life" problems into algebraic equations in order to solve them. We will do this now with literal equations [or formulas if you like]

During a storm, you can estimate **the number of miles away** $[m]$ a bolt of lightning strikes by first counting the number of seconds $[s]$ between the bolt of lightning and the associated clap of thunder, and then dividing by 5.

We can create a **literal equation** [formula] with this information. We are interested in how far we are from the storm which in this case is m [miles].

m [number of miles] = s [number of seconds between the lightning and thunder] divided by 5

or: $\qquad m = \dfrac{s}{5}$

Now suppose that we wanted "solve" this literal equation for seconds $[s]$ in terms of miles $[m]$. We need to find a way of getting s **alone** on one side of the equation.

Right now s or $1s$ is being divided by 5: $\qquad m = \dfrac{1s}{5}$

We can eliminate the 5 in the denominator **[right side of equation]** by multiplying both sides of the equation by 5 **[or $\dfrac{5}{1}$ which is equivalent to 5]**: $\qquad 5 \cdot m = \dfrac{1s}{5} \cdot \dfrac{5}{1}$

Notice that on the right side, we have $\dfrac{1}{5}$ and its **reciprocal**, $\dfrac{5}{1}$ multiplying each other, which creates the **identity element**, 1, for multiplication and can be eliminated. As we have previously mentioned, this is the principle behind "canceling": $\qquad 5 \cdot m = \dfrac{1s}{\cancel{5}} \cdot \dfrac{\cancel{5}}{1}$

Which results in: $\qquad 5m = \dfrac{1s}{1}$

Since $\dfrac{s}{1} = s$, we can re-write the right side of the equation as: $\qquad 5m = 1s$

Now we interchange the right and left sides of the equation since they are equivalent: $\qquad 1s = 5m$

or its equivalent: $\qquad s = 5m$

We have solved the equation for s in terms on m.

Concept Homework

As an assessment of your understanding of the concepts set forth in this section, answer the following questions. If necessary, review the material to help you arrive at the correct conclusions.

1. Explain what it means to solve a literal equation.

2. A formula is fundamentally different than a single variable equation. Explain the difference and make up examples of the two not used in this topic.

3. What does a literal equation tell us about its variables and how is it limited in a way that a single variable equation is not?

4. How is the strategy the same for solving both literal equations and one-variable equations?

5. Make up a literal equation and show the steps for solving it for a different variable contained therein.

Exercises

1. Solve for x in terms of y :

    a. $3x + y = 9$ **b.** $4x + 3y - 12 = 0$

2. Solve for y in terms of x :

    a. $3x + y = 9$ **b.** $4x + 3y - 12 = 0$

3. If $A = P + Prt$

    a. Solve for r in terms of A, P and t

    b. Solve for t in terms of A, P and r

Solutions: 1a. $x = 3 - \frac{1}{3}y$ or $x = \frac{9-y}{3}$; 1b. $x = \frac{-3y+12}{4}$ or $x = -\frac{3}{4}y + 3$; 2a. $y = -3x + 9$; 2b. $y = \frac{-4x+12}{3}$ or $y = -\frac{4}{3}x + 4$;

3a. $r = \frac{A\text{-}P}{Pt}$; 3b. $t = \frac{A\text{-}P}{Pr}$.

Topic 10 - Inequalities: How They Differ from Equations

Equations deal with two equivalent quantities connected with an "equals" [=] symbol.

Besides utilizing equations, it is also useful when solving some problems to consider the relationship between quantities that are unequal. In these cases quantities are connected by **inequality symbols** $\left[>,\ \geq,\ <,\ \leq\right]$.

These symbols represent the words:

> "greater than"
$\geq$ "greater than or equal to"
$<$ "less than"
$\leq$ "less than or equal to"

Equations and inequalities are similar in many ways and can be manipulated using many of the same properties. However, there are some **significant differences** between them which we will discuss in this topic.

Here are **two** important characteristics of inequalities that **differ** from equations:

1. There is only **one solution** to a **conditional** linear equation in one variable to the first power while **an inequality has an *infinite* number of solutions**.

   [Note: A *solution* is a value for the variable that makes an equation *or* inequality true.]

Example 1.

First we will examine an equation.   Solve for x: $x + 2 = 3$

We would solve this equation by adding -2 to both sides of the equation:
$$\begin{array}{r} x + 2 = 3 \\ -2 \quad -2 \end{array}$$

thereby getting x alone on one side of the equation.
Adding **opposites** on the left side of the equation results in 0, so: $x = 1$
[solution]

$x = 1$ is the **unique** solution. That is, $x = 1$ is the **only** value for x that will make the equation true. Now let's **change the equation to an inequality.**

Solve the **inequality:** $x + 2 \leq 3$.
[$\leq$ is a symbol that means "less than or equal to.]
We would solve this inequality using the same strategy.

Add -2 to both sides of the inequality [just as we did with the equation]:
$$\begin{array}{r} x + 2 \leq 3 \\ -2 \quad -2 \end{array}$$

The ***addition property of equations*** also holds true for inequalities so: $x + 0 \leq 1$
eliminating the identity element, 0, for addition: $x \leq 1$
" x is less than or equal to 1 " is ***solution set*** because there is more than   [solution]
one solution.

Note: "a solution" to either an equation or an inequality is **any value for the variable that will make them true**.

We can substitute any value for x that is **either less than or equal to** 1 and it will make the original inequality, $x + 2 \leq 3$, true. Let's substitute 0 for x in: $\quad x + 2 \leq 3$

This results in: $\quad 0 + 2 \leq 3$

As can be seen, this is a **true statement**. We could have used **many other values** for x since there are many other numbers less than or equal to 1. [1, −1, −2.5, −100.75, etc.]

Any of these numbers would make the inequality, $x + 2 \leq 3$, true. Think of **any** number less than or equal to 1 and substitute into the inequality. The result will be a true statement. Therefore, the inequality has an **infinite** number of solutions.

2. When we manipulate inequalities, we **use the same properties that we do with equations _with the following exceptions_** :

Exception #1: Interchanging the Left and Right Sides of Inequalities.

First, lets look at the **equation** $x = 5$. It is equivalent to $5 = x$. **[The left side equals the right side.**
Therefore, they are interchangeable. Ex: if $2 + 3 = 5$, then $5 = 2 + 3$ **]**

Or, if we had the equation, $3x + 2 \;=\; 5x - 3$, it would be equivalent to $5x - 3 \;=\; 3x + 2$

left side$\quad$right side $\hspace{4cm}$ [right side interchanged with left side]

In other words, when we **interchange** the left and right sides of any equation, **the equation is still valid.**

This is *not* **true with inequalities!**

For example, $\quad x < 1 \quad$ is **_not_** equivalent to $\quad 1 < x$.

" x is less than 1 " $\hspace{5cm}$ " 1 is less than x "

But $\qquad x < 1$ *is* equivalent to $1 > x$

Notice that the _inequality symbol must be reversed_ when you interchange the left side with the right side of an inequality.

Why? It is easier to see why when the inequality statement has no variables: $\qquad 6 < 8$

6 **is** less than 8, so this is a true statement.

Now let's interchange the left and right side of the inequality: $\qquad 8 \not< 6$

This is no longer a true statement! [8 is **not** less than 6]

We must *reverse* the inequality symbol, to make it true: $\qquad 8 > 6$

Now it is a true statement.

We can make up a number of examples without variables and it will become evident that **this is always true.** We can then utilize this "rule" when variables are involved.

We should **not** leave a solution to an inequality with the variable on the right side. The variable should always be on the left side. **[We will see later on that this is necessary for graphing inequalities.]**

For example: If we get a solution to an inequality: $-3 \le y$

we should rewrite this with its **equivalent**: $y \ge -3$

Exception #2: Dividing or Multiplying by a Negative Number

We know that the **multiplication and division property of equations** allows us to multiply or divide both sides of an **equation** by any quantity, either positive or negative and still have a valid equation. However, **when we divide or multiply both sides of an inequality *by a negative number*,** we must *reverse the inequality sign.*

Example 1.   [Let's start with an inequality that doesn't have variables.]

Consider the following true statement: $-3 < 6$

The division property of equations **does not hold true** for inequalities when we **divide by a negative number!**

Therefore, when we **divide both sides of the inequality by** -3 : $\dfrac{-3}{-3} > \dfrac{6}{-3}$

the inequality sign **must be reversed to make the statement true!**

When the fractions are simplified the result is: $1 > -2$

This is a **true** statement, only because we **reversed the inequality symbol.**

Dividing by a **negative *divisor*** $\begin{bmatrix}-3\end{bmatrix}$ required that we reverse the inequality sign from $<$ to $>$.

Example 2.   [Now, let's look at an inequality with a variable.]

Solve for x : $6 \le -2x$

Just as with equations, to solve for x we must manipulate the inequality so that x has a coefficient of 1. Therefore we must divide both sides of the inequality by -2. When we do this, we must **reverse the inequality symbol**

as shown: $\dfrac{6}{-2} \ge \dfrac{-2x}{-2}$

Resulting in: $-3 \ge 1x$

This solution should have the variable on the left side, so we interchange the left side and right side by again **reversing the inequality symbol:** $1x \leq -3$

Or just: $x \leq -3$

[solution]

This solution means that whenever x is less than or equal to -3, the inequality, $6 \leq -2x$, will be true!

Example 3. [Again, let's start with an inequality that doesn't have variables.]

Consider the following **true** inequality: $4 \geq 2$

Let's see what happens when we **multiply** both sides by -2.

The multiplication property of **equations** does **not** hold true for inequalities when we **multiply be a negative number:** $(-2)(4) \not\geq (-2)(2)$

Performing the multiplication results in: $-8 \not\geq -4$

We see that the inequality is no longer true

[-8 is not greater than or equal to -4]

We must **reverse the inequality sign** to make it true: $-8 \leq -4$

[-8 **is** less than or equal to -4]

This is now a true statement!

We must always reverse an inequality sign when multiplying (or dividing) by a negative number.

Example 4. [We will now consider an inequality with variables.]

Solve for x : $-1x < 9$

This inequality is not yet solved because x is not positive.

To make x positive, we can multiply both sides of the equation by -1 and, therefore, **must reverse the inequality symbol**: $(-1)(-1x) > (-1)(9)$

Simplifying the inequality, we get: $1x > -9$

Or its equivalent: $x > -9$

[solution]

For any value for x that is greater than -9, the inequality will be true. Therefore, there are an *infinite* number of solutions to this inequality.

For example, if we substitute 10 for x in the **original inequality** [$-1x < 9$]: $-1(10) < 9$

Simplifying: $-10 < 9$

This is a **true statement.**

Other possible solutions could be: $x = -8$, $x = -7$, $x = -6$, etc. since, when substituted for x in the inequality, $-1x < 9$, the statement would be true. As can be seen, there are an **infinite number of solutions** to this inequality since we can substitute an infinite number of numbers for x that are greater than -9. We can extend this logic to say that **all inequalities have an infinite number of solutions.**

TWO MORE IMPORTANT THINGS TO CLARIFY:

1. Although a negative *divisor or multiplier* causes the inequality symbol to be reversed, the sign of the number *being divided* or *multiplied* does *not* affect the inequality symbol.

> **Example 1.**   Let's start with the true inequality: $-3 > -8$
> Multiply both sides of the following inequality by 2: $(2)(-3) > (2)(-8)$
> When we perform the multiplication, we get: $-6 > -16$
> which is a **true statement.**

Even though the numbers in the inequality were negative, We did **not reverse the inequality symbol** because the *multiplier* [2] was **positive** .

> **Example 2.**   **Solve for *x*:** $2x < -16$
> To solve, we divide both sides by 2: $\dfrac{2x}{2} < \dfrac{-16}{2}$

Notice that **the inequality symbol did *not* change** even though the number on the right side of the inequality [–16] is negative.

> Performing the division, we get: $1x < -8$

We did **not reverse** the inequality sign because *the divisor,* 2, was **positive**. The solution is: $x < -8$

We can check the solution to $2x < -16$ by substituting some number that is less than -8 for x in the original inequality.

> Let's use -10: $2(-10) < -16$
> Simplifying, we get: $-20 < -16$
> Which is a **true statement.**

You can use **an infinite** number of numbers that are less than -8 to further verify that the solution is true.

2. **Addition and subtraction** operations <u>**do not**</u> affect the inequality symbol, **regardless of whether we are adding or subtracting negative numbers.**

> **Example 1.**   **Solve for *x*:** $2x + 3 \geq 11$
> $$2x + 3 \geq 11$$
> Adding -3 to both sides:
> $$\underline{ -3 \quad -3}$$

Since the opposites, 3 and -3 equal 0 on the left side of the inequality: $2x \quad \geq 8$

Notice that the **inequality sign didn't change** because we were **adding** a negative number, not **multiplying** or **dividing** by it. .

$$\text{Dividing both sides } 2: \qquad \frac{2x}{2} \geq \frac{8}{2}$$
$$\text{This results in:} \qquad x \geq 4$$

[solution]

You can check the solution in the original inequality, $2x+3 \geq 11$, with x being any number greater or equal to 4. It will make the inequality, $2x+3 \geq 11$, a true statement.

Graphing Inequalities

The **solution sets** of inequalities are all quantities that make and inequality true. These solutions can be represented visually by **graphing** them on a **number line** or described by **interval notation**. Here are some examples:

Example 1. Graph the **solution set** of $x > 3$

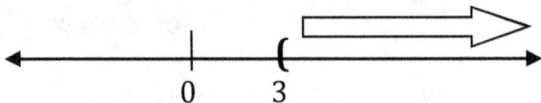

$$0 \qquad 3$$

Interval Notation: $(3, \infty)$

The left boundary of the inequality is the first number in interval notation and the right boundary is the second number. In this case, numbers greater than 3 go on **forever**. Therefore, we use the "infinity" symbol for the right boundary.

Notice that 3 is not included in the solution set $x > 3$ **[3 does not make the inequality true]**. A parenthesis symbol, " (" , implies that 3 is not included. The infinity symbol, ∞, is always marked by a parenthesis symbol, ") " , since we can never get to it, so it can't be included.
On the graph, we use the same symbol , "**(**", to indicate that 3 is not included in the solution set.

Example 2.   Graph the **solution set** of $x \leq 1$

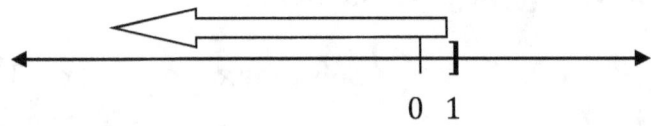

$$0 \ 1$$

Interval Notation: $\left(-\infty, 1\right]$

The left boundary of the inequality is the negative infinity symbol, $-\infty$. In this case, the numbers 1 or less go on forever in the leftward direction. A parenthesis symbol, " (" , implies that $-\infty$ is not included in the inequality since we can never get to it. Notice that 1 **is** included in the solution set $x \leq 1$, since if x takes on the value of 1, the statement is true [$1 \leq 1$ **is a true statement**]. The bracket symbol, "**]**" is used for the right boundary to indicate that the right boundary, 1, is included in the inequality. On the graph, we use the same symbol, "**]**", to indicate that 3 **is** included in the solution set.

Example 3. [Suppose we wished to graph an inequality that exists between **two distinct points**.]

Graph the **solution set** of $-4 < x < 5$

This notation is telling us that x is greater than -4 [$-4 < x \Rightarrow x > -4$] and less than 5. That means that the inequality lies **between** -4 and 5. It would be graphed as follows:

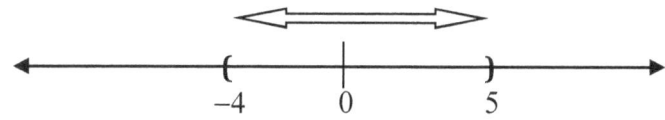

Interval Notation: $\left(-4, 5\right)$

Application

The perimeter of a rectangle can be no greater than 90 inches and the width must be 20 inches. What is the maximum length of the rectangle? [$2l + 2w = P$]

The problem states that the perimeter can be **no greater than** 90 inches. This implies an inequality. "No greater than" translates mathematically into "less than or equal to". Therefore we must change the formula for perimeter to the inequality: $2l + 2w \leq P$. Since the width in this case is fixed at 20 inches, and the perimeter is given a maximum of 90 inches, we can mathematically translate the problem to an inequality in one variable:

$$2l + 2 \cdot 20 \leq 90$$

Performing the multiplication on the left side of the inequality: $2l + 40 \leq 90$

Adding -40 to both sides of the inequality: $-40 \quad -40$

Results in: $2l \qquad \leq 50$

Dividing boths sides of the inequality by 2: $\dfrac{2l}{2} \qquad \leq \dfrac{50}{2}$

Results in: $l \leq 25$

Therefore, **The maximum length of the rectangle** [must be less than or equal to] **is 25 inches.**

Concept Homework

As an assessment of your understanding of the concepts set forth in this section, answer the following questions. If necessary, review the material to help you arrive at the correct conclusions. With "true or false" statements, if the statement is true, make up an example that supports the statement without using examples given in the material. If a statement is false, correct the wording to make it true. Then make up an example that supports the statement without using examples given in the material.

1. True or False:

    a. We follow the same strategy for solving inequalities as we do for equations in that we isolate the variable and makes its coefficient 1.

    b. The addition property of equations holds true for inequalities, regardless of whether we are adding or subtracting positive or negative numbers.

    c. If $2x \leq -8$, then $x \geq -4$. Explain your answer.

    d. If $-3x \geq 9$, then $x \geq -3$. Explain your answer.

    e. If $x < 11$, then $11 < x$. Explain your answer.

    f. $-4 \geq -4$ is a true statement. Explain your answer.

    g. $-4 < -4$ is a false statement. Explain your answer.

    h. There are an infinite number values for x that will make $3x = 15$ true. Explain your answer.

    i. There are an infinite number values for x that will make $3x < 15$ true. Explain your answer.

Solutions: a. True; b. True; c. False; d. False; e. False; f. True; g. True; h. False; i. True

2. Give three examples of manipulating inequalities where the inequality sign is reversed, using all the exceptions outlined in this topic . Make up original examples.

3. Describe in words the solution sets for the following inequalities:

    a. $x \geq -2$ b. $x < -6$ c. $-7 \leq x < 9$

Exercises

Solve the following inequalities. Graph each solution and describe it in interval notation.

1. $x-4>5$ **2.** $12-2x\le16-3x$ **3.** $10x-1-2x>7x-6$ **4.** $6-7(2-3x)\le5(1+4x)$

5. $7-4(x-3)<3(7x-2)$ **6.** $7x-8\ge10x+4$ **7.** $5x+5<6x-5$ **8.** $\dfrac{5}{12}x-\dfrac{3}{4}x>-3$

9. Joe wants to rent a car the weekend **[2 days]** to go traveling. He has budgeted no more than $110.00 for the cost of the car rental. The rental company charges $30 per day and $.20 per mile traveled. What is the maximum number of miles that Joe can travel and stay within his budget??

Solutions: 1. $x>9$ $(9,\infty)$; 2. $x\le4$ $(-\infty,4]$; 3. $x>-5$ $(-5,\infty)$; 4. $x\le13$ $(-\infty,13]$; 5. $x>1$ $(1,\infty)$; 6. $x\le-4$ $(-\infty,-4]$; 7. $x>10$ $(10,\infty)$; 8. $x<9$ $(-\infty,9)$ 9. $x\le250$ miles.

Topic 11 – The Relationship Between Linear Equations and Their Graphs

Locating Ordered Pairs on the Coordinate System

In order to locate a point on the coordinate system, we need **two measurements** which together are called **ordered pairs**. The first value of the ordered pair (the x - value) determines how far **left [negative numbers]** or **right [positive numbers]** the point is located from the $y-axis$. The second value of the ordered pair (the y - value) determines how far **up [positive numbers]** or **down [negative numbers]** the point is located from the $x-axis$. In order to locate a point on the coordinate system, **we need both of these measurements** combined. Below is a coordinate system with the locations of various ordered pairs:

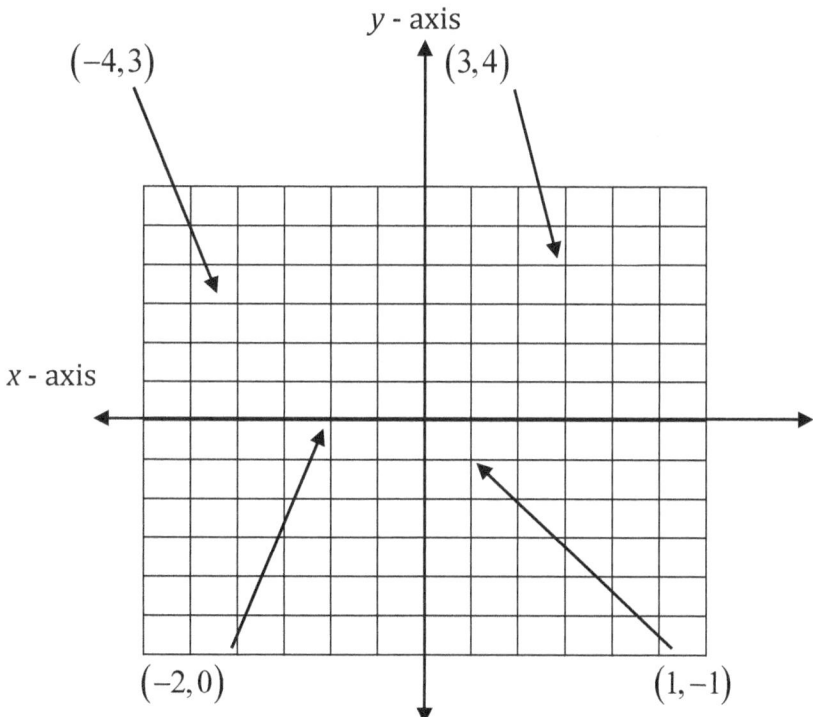

Linear Equations

A **linear equation** [in all but two special cases which we discuss later in this topic] has **two** variables that we generally name x and y that are **to a power of 1.**

Finding Solutions to Linear Equations.

In order to find a solution to such an equation, we must **arbitrarily assign** [we pick any value we like!] **any** real value to one of the variables, which will allow us to find the corresponding value for the other variable. These values for the two variables **will make the equation true**.

Example: In the linear equation $2x+y=8$, to find a *solution* to the equation [values for x and y that will make the equation true], we **arbitrarily** assign a value to *either x* or *y*.

Let's consider the equation: $2x + y = 8$

Suppose we assign the value 2 for y [it could have been any number!].

We say "let $y = 2$". Substituting 2 for y: $2x + 2 = 8$

Notice that we now have an equation with *only one variable* so we will be able to *solve it for x*!

Adding -2 to both sides of the equation:
$$2x + 2 = 8$$
$$\underline{-2 \quad -2}$$

Results in: $2x + 0 = 6$

Dividing both sides of the equation by 2: $\dfrac{2x}{2} = \dfrac{6}{2}$

We have now created the coefficient of 1, $\left[\dfrac{2}{2}\right]$ for x: $1x = 3$

Or just: $x = 3$
[solution]

We have found the <u>corresponding</u> value for *x* $\left[x = 3\right]$ that solves the equation when <u>we chose</u> a value for *y*. $\left[\text{Let } y = 2\right]$. Therefore, *one* solution [of many] **to this equation is $(3, 2)$. This is called an *ordered pair* where 3 is the *x* value and 2 is the *y* value. These values will make the equation** $2x + y = 8$ **true.**

Let's find **another solution** to this equation. This time let's **arbitrarily** assign a value to x.
[We can pick any value we like.]:

Arbitrarily let $x = 4$ in: $2x + y = 8$

Assigning x the value of 4: $2(4) + y = 8$

Simplifying: $8 + y = 8$

Adding -8 to both sides of the equation: $\underline{-8 \qquad -8}$

Results in: $0 + y = 0$

or just : $y = 0$
[solution]

We have found the corresponding value for *y* [y = 0] **that makes the equation true when the value of *x* is** 4 [Let x = 4]**! Therefore, *another* solution** [of many] **to this equation is $(4, 0)$. This is another ordered pair where 4 is the *x* value and 0 is the *y* value.**

Why would there be **many** solutions to this equation? Because we could **arbitrarily** choose many values for **y** other than 2 or for **x** other than 4 and each time find many corresponding values that will make the equation true. We can arbitrarily assign an **unlimited number of values** to <u>either</u> **x** or **y**! In these **ordered pairs**, the x value is always listed first in the parenthesis [indicating an ordered pair] and the y value is always listed second. The coordinate system lends itself to visualizing solutions to linear equations. Locating a point requires two values as do solutions to linear equations!

<u>Now</u> <u>here is an amazing fact</u>! If we plot *every* solution to a particular linear equation on a coordinate system, the collection of these solution points [ordered pairs] will **form a straight line!** That is why these equations are called <u>linear</u> **equations.**

> If we plot these two solution points that we found earlier on a coordinate system and draw a straight line through them, this line will graphically represent *every* ordered pair that will make the equation $2x + y = 8$ true!

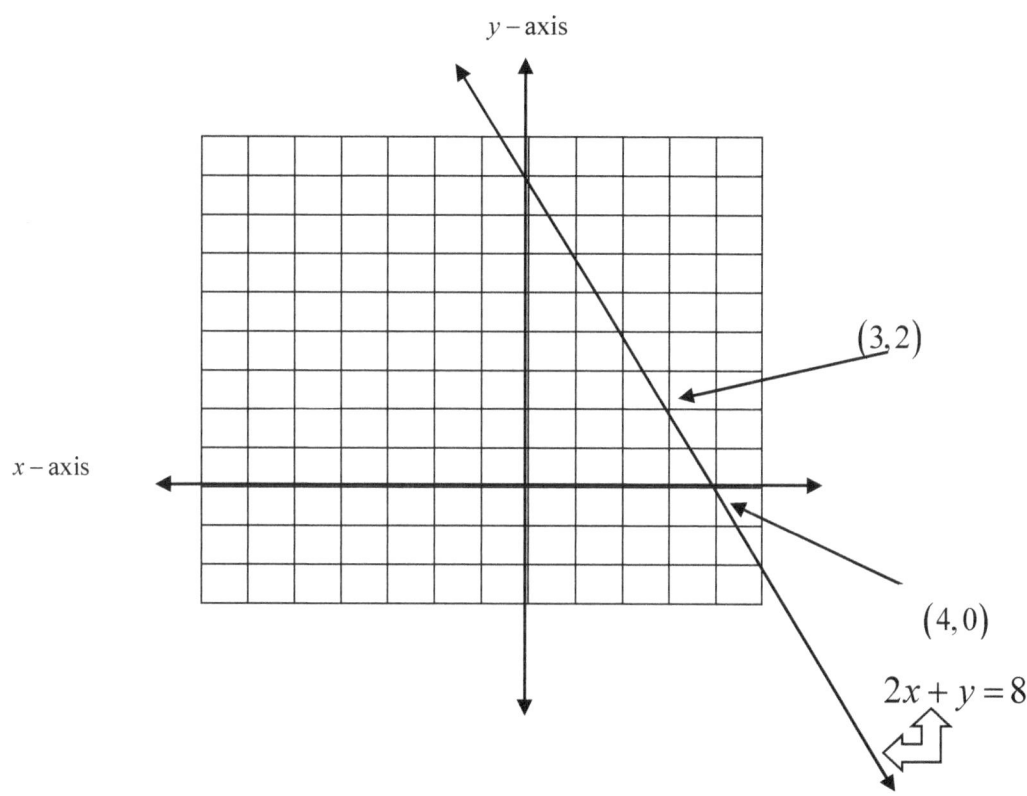

y – axis

x – axis

$(3, 2)$

$(4, 0)$

$2x + y = 8$

<u>Any</u> points on this line [$(0,8);(1,6);(1.5,5)(2,4);(5,-2)$], just to name a few [or any points between or beyond them on the line], when put into the equation $2x + y = 8$, will make this equation true!

Special Linear Equations

There are two types of **special linear equations.** You should **practice how these special equations are graphed. These equations are special because they only have one variable!**

They are: (1) $x = c$ [where c is *any* <u>constant</u> (stand-alone number)]
                (2) $y = c$ [where c is *any* <u>constant</u> (stand-alone number)]

Let's take the first case and ***arbitrarily*** substitute 4 for c, making the **special equation:** $x = 4$.

     [What does "arbitrarily" mean in this context? I could have chosen *any* real number.]

What do we do about y values for this equation?

Well, let's add a y variable ***without really changing the equation***! We do this by multiplying y by 0 ! Then the equation would look like: $\quad x+0y=4$

[Since $0y$ is $0 \cdot y$ which has a value of 0, the balance of the equation has not changed. $0 \cdot$ *any* number = 0]

Now let's **arbitrarily** assign -2 to the **y** variable: $\quad x+0 \cdot (-2) = 4$

performing the multiplication: $\quad x + 0 = 4$

removing the identity element for addition : $\quad x = 4$

So we see that if $y = -2$, then $x = 4$, so one solution to the equation $x = 4$ is: $\quad (4, -2)$

Let's get a second solution: $\quad x + 0y = 4$

We arbitrarily assign 5 to the y variable: $\quad x + 0 \cdot 5 = 4$

Performing the multiplication: $\quad x + 0 = 4$

Removing the identity element for addition: $\quad x = 4$

So we see that if $y = 5$, then $x = 4$, so another solution to the equation $x = 4$ is: $\quad (4, 5)$

In Fact, **no matter what value we assign to y, x will remain 4 since the y value keeps getting multiplied by 0.** Here are some other solutions: $(4, 0)$, $(4, 1)$, $(4, -3)$ that will all be on the same solution line. We will plot the first two solutions on the coordinate system below. By drawing a line through these points, they will determine the solution line [a collection of *all* the solutions] for $x = 4$:

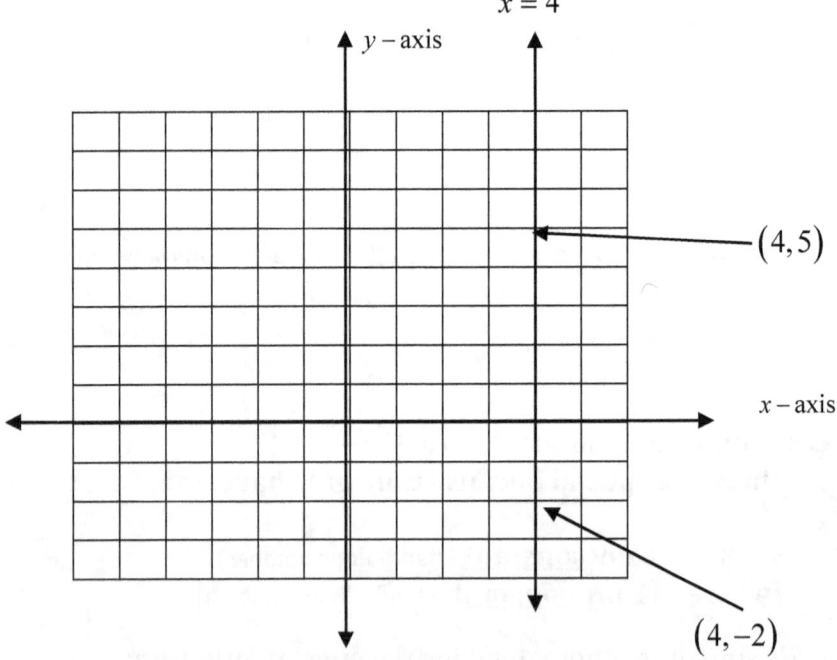

You can see that if we draw the line containing *all* of the solutions, it is a ***vertical line*** ["straight up and down"] ***where every x - value is*** 4 !

Now let's look at the **second special equation** :

$y = c$ [where c is any **constant** (stand alone number)]

We will *arbitrarily* make the second equation: $y = 3$

What do we do about x values for this equation?

Let's add an x variable without changing the **balance** of the equation: $0x + y = 3$

Since the coefficient of x is 0, the equation has not really changed!

Now let's *arbitrarily assign* -2 to the x variable: $0 \cdot (-2) + y = 3$

performing the multiplication: $0 + y = 3$

removing the identity element for addition: $y = 3$

so we see that if $x = -2$, then $y = 3$. This is **one** solution to the equation: $(-2, 3)$

Let's get a **second solution** by arbitrarily assigning 5 to the x variable: $0 \cdot 5 + y = 3$

performing the multiplication: $0 + y = 3$

removing the identity element for addition: $y = 3$

so when $x = 5$, $y = 3$, **another** solution to the equation is: $(5, 3)$

In Fact, **no matter what value we assign to x, y will remain 3 since the x value keeps getting multiplied by 0.**

Here are some other solutions: $(0,3)$, $(1,3)$, $(-3,3)$. They all lie on the solution line. We will plot the first two solutions on the coordinate system below. By drawing a line through these points, they will determine the solution line **[a collection of *all* the solutions]** for $y = 3$.

$(-2,3)$ y - axis $(5,3))$

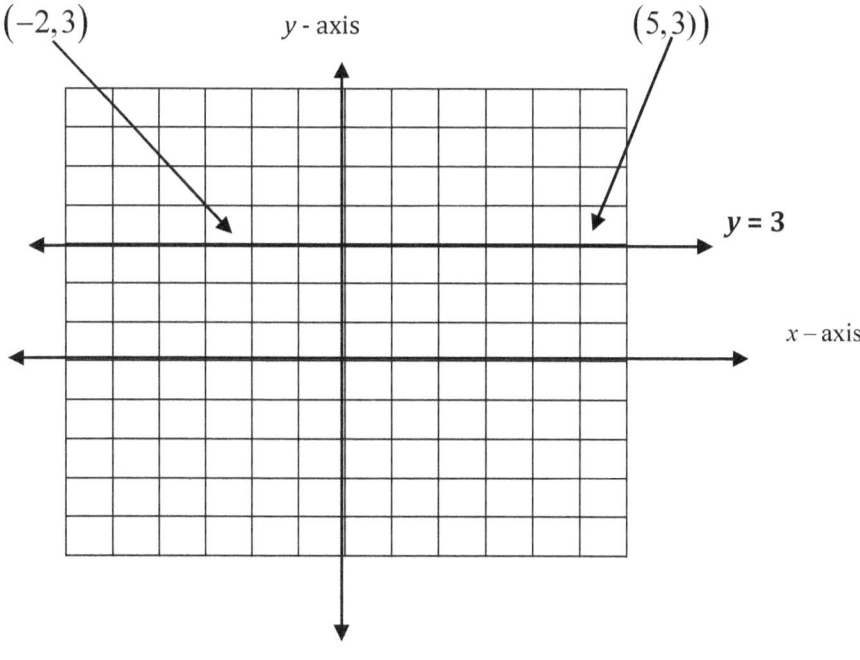

$y = 3$

x – axis

You can see that the y – values for *all* these solutions, is 3 and the line containing all the solutions is a *horizontal line!*

Summary for special linear equations with just one variable:

$x = c$, where c is <u>any constant</u>, will always be a ***vertical*** line [up and down] where every x value in all ordered pair solutions is c, the <u>constant</u> in the equation.

$y = c$, where c is <u>any constant</u>, will always be a *horizontal* line [flat] where every y value in all ordered pair solutions is c, the <u>constant</u> in the equation.]

Concept Homework

As an assessment of your understanding of the concepts set forth in this section, answer the following questions. If necessary, review the material to help you arrive at the correct conclusions. With "true or false" statements, if the statement is true, make up an example that supports the statement without using examples given in the material. If a statement is false, correct the wording to make it true. Then make up an example that supports the statement without using examples given in the material.

A. True or False

1. We can't find a solution to a linear equation in two variables unless we assign a value to one of the variables.

2. We should always assign a value to the x variable to find a solution for the y variable. If we assign a value to the y variable, we cannot be sure that we will get a correct ordered pair solution.

3. There is only one solution to any linear equation in two variables.

4. When we plug in the solution [ordered pair] to a linear equation, it will make the equation true.

5. Some linear equations have only one variable.

6. In the equation $x = -3$, whatever values we choose for y, the resulting x values will always be -3.

7. $x - 3 = 0$ is a linear equation whose graph will be a vertical line.

Solutions: True; 2. False; 3. False; 4. True; 5. True; 6. True; 7. True.

Exercises

Use the graph paper below to executive the following problem:

B. Make up an equation with only a y variable. Find three solutions for the equation and plot them on a coordinate system.

C. Make up an equation with only an x variable. Find three solutions for the equation and plot them on a coordinate system.

D. Show on a coordinate system where $y = 0$ can be found.

E. Create *any* two-variable equation without zero coefficients **[not used as an example in these notes]**.

Graph the equation by finding two solutions.

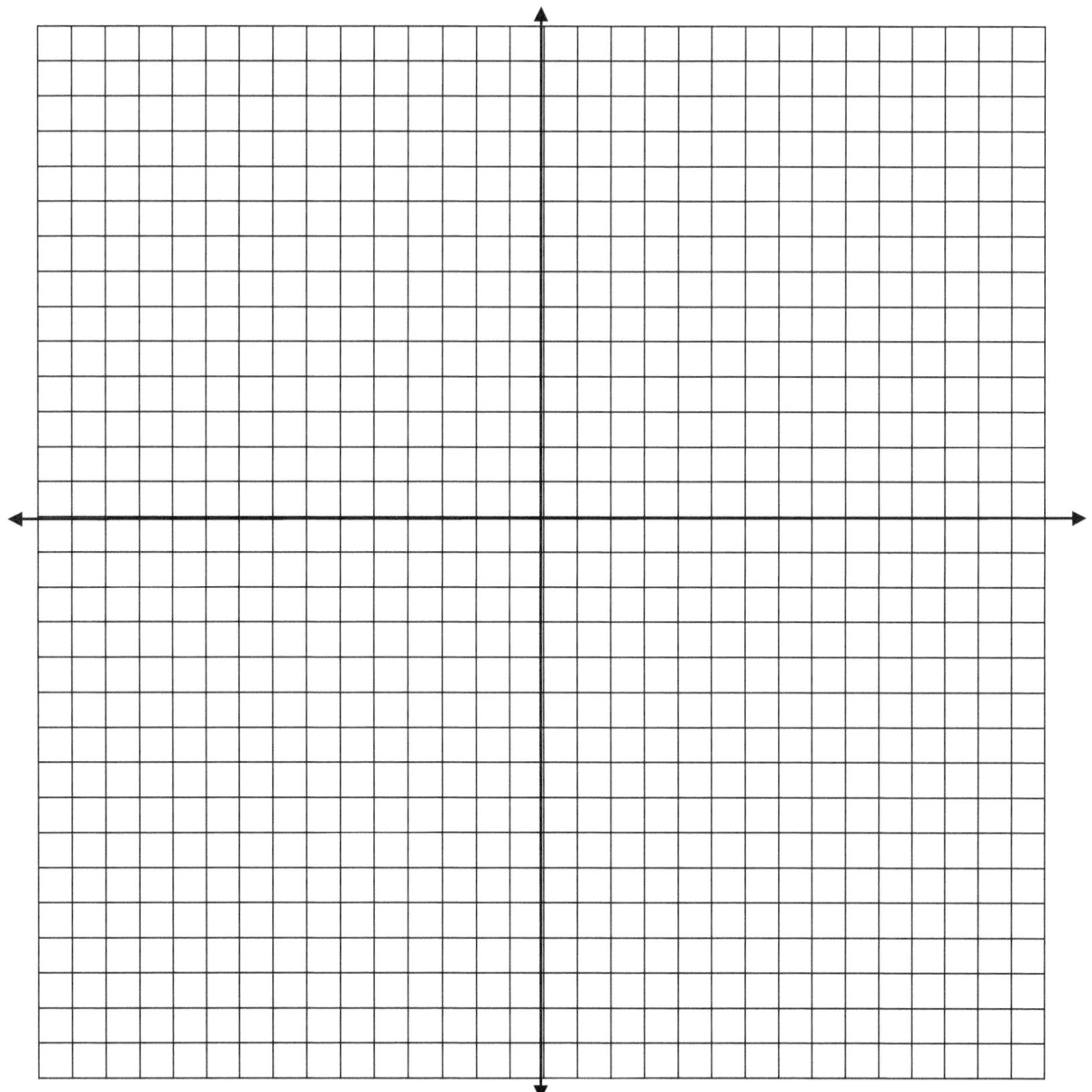

Topic 12 - The Concepts of x and y Intercepts, Slope, and the $y = mx + b$ Form of Linear Equations

With the exception of horizontal and vertical lines, any straight line representing the solution set of some linear equation must cross **both** the x-axis and y-axis of the coordinate system. If you envision a cooridinate grid that goes on forever and superimpose **any** line [except horizontal or vertical lines] on it that extends forever in each direction you will see that this is true. Now envision a horizontal or vertical line. Does either one cross **both** the x- and y – axes? [No. Only one axis]. These crossing points are called **the x- and y-intercepts** of that solution line. Because a line is straight, its **steepness** will be consistent throughout the line. In other words, the steepness never varies, regardless of segment of the line that we observe.

The amount of steepness can be measured mathematically and is called **the slope** of the line. A line may be differentiated from **all other lines** by determining two of its characteristics: its slope and its y-intercept. As we will see in this section, we can manipulate any linear equation in such a way that its slope and y-intercept can easily be determined. **We can then use these characteristics to graph its solution line .**

Linear Equations can **model real-life situations** and be used to predict the outcome for one variable as the value for the other variable changes.

The x and y Intercepts of a Linear Equation

In the last topic, we discussed how we could find solutions to a linear equation and thereby graph the equation's solution line. In order to do this, we **arbitrarily** assigned *any* real value to one of the variables, [either x **or** y], which allowed us to find the corresponding [x or y] value for the other variable that will make the equation true.

Let's consider the equation: $x + 2y = 4$

If we **arbitrarily** assign a value of 0 to the x variable [We say "Let $x = 0$ "]

then substituting 0 for x: $0 + 2y = 4$

Notice the equation **now** only has one variable, so it can be solved!
Since the y term should have a coefficient of 1 ,we divide both sides of

the equation by 2 : $\dfrac{2y}{2} = \dfrac{4}{2}$

Now the coefficient of y is $\left[\frac{2}{2} = 1\right] 1$, resulting in: $1y = 2$

So, we find that, when $x = 0$, the corresponding y value that makes the

equation true is: $y = 2$

This tells us that the equation, $x + 2y = 4$, **is true** when $x = 0$ and $y = 2$.

As an ***ordered pair***, this solution to $x + 2y = 4$ is written: $(0,2)$

Let's graph this solution as a point on a coordinate system:

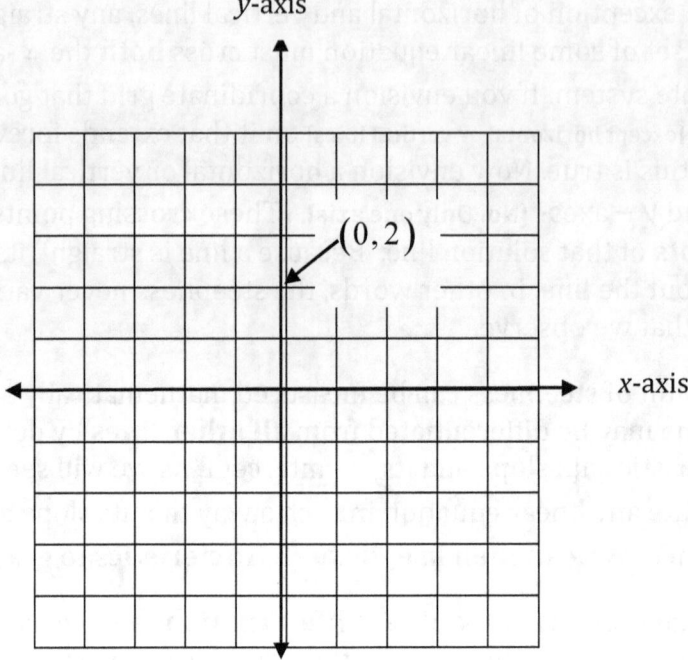

Notice that this point lies on the $y-axis$! **This will always be the case when we arbitrarily assign the value of** 0 **to the** x **variable!** Why is this so? Think about **every point** on the $y-axis$. **They all have** 0 **for their** x **values!** [Pick out *any* point on the $y-axis$. **What is its** x **value**?] Therefore, when we assign 0 to the x variable, the solution point for that equation **will have to be on the** $y-axis$!

Now, with the same equation: $\qquad x + 2y = 4$

Let's **arbitrarily** assign a value of 0 to the y variable [we say "**Let** $y = 0$ "]

Then substituting 0 for y : $\qquad x + 2(0) = 4$

Notice the equation **now** only has one variable, so it can be solved!

since multiplying *any* number by 0 will result in 0 : $\qquad x + 0 = 4$

Simplifying: $\qquad x = 4$

This tells us that the equation, $x + 2y = 4$, is **also true** when $x = 4$

and $y = 0$. As an ***ordered pair***, this solution to $x + 2y = 4$ is written: $\qquad (4, 0)$

Let's graph both $(0,2)$ and $(4,0)$ on a coordinate system:

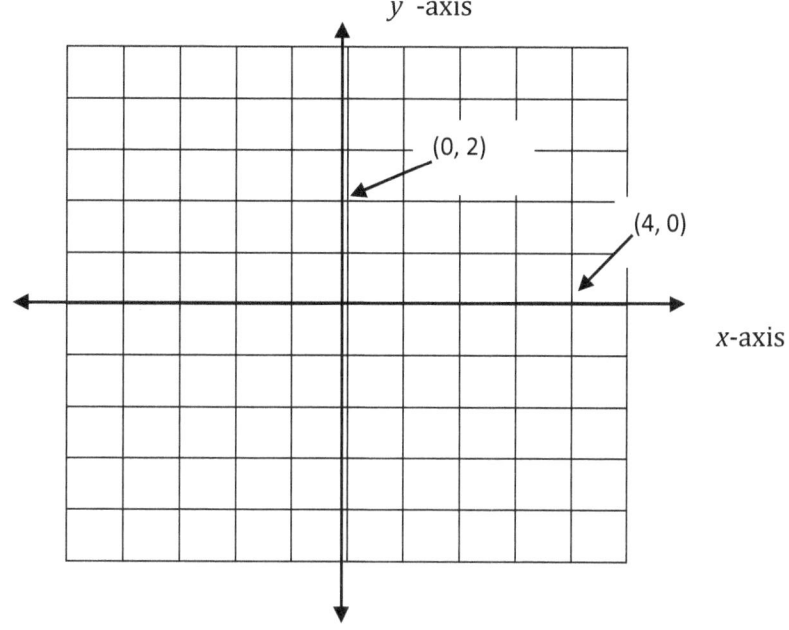

Notice that $(4,0)$ lies on the $x-axis$ **! This will always be the case when we arbitrarily assign the value of** 0 **to the** y **variable!** Why is this so? Think about **every point** on the $x-axis$. **They _all_ have** 0 **_for their_** y **_values!_** [Pick out _any_ point on the $x-axis$. **What is its** y **value?**] Therefore, **when we assign** 0 **to the** y **variable, the solution point for the equation will have to be on the** $x-axis$ **!**

With only these two solution points, we can draw the entire solution line for the equation $x+2y=4$, since linear equations **always** have **straight** solution lines. As we learned in the last topic, every solution to a particular linear equation will lie on a unique straight line. We show this solution line below:

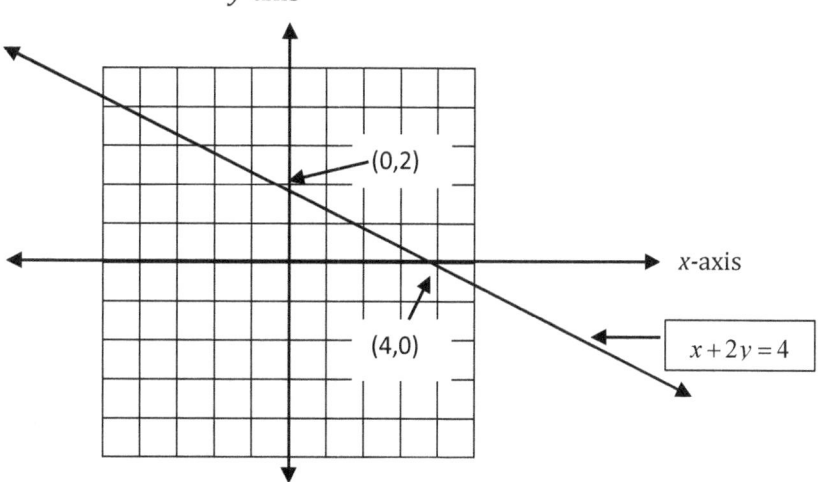

As can be seen, the point $(0,2)$ is where the solution line **crosses** the $y-axis$. We call such a point, the y-**intercept** of a solution line. We can also see that the point $(4,0)$ is where the solution line **crosses** the $x-axis$. We call such a point the x-**intercept** of a solution line.

We can **always** find the y and x **intercepts** of **any** solution line in this manner:

y-**intercept:** [Let $x=0$] x-**intercept:** [Let $y=0$]

The Slope of a Solution Line

Because the "solution line" of a linear equation [the set of all the solutions] is straight, its "steepness" is **constant.** That is, it is the **same anywhere on the line**. The mathematical name for the steepness is the line's "**slope**" and is denoted mathematically by the lower case letter m. Mathematically, the slope of a line is the ratio of the **difference** between the y measurements and the x measurements of **any** **two points** on the solution line. This slope, or rate of change can be determined with the formula $m = \dfrac{y_2 - y_1}{x_2 - x_1}$ where (x_1, y_1) represents **any** solution [ordered pair] **on the solution line** and (x_2, y_2) represents **any** **other** solution **on the solution line**. This can be thought of as **the rate of change** of the y and x values on the solution line. How do we choose which points on the solution line to assign to (x_1, y_1) and (x_2, y_2)? **It doesn't matter**, as long as the points chosen are **on** the solution line! **We can choose _any_ two solution points for the equation to determine its slope. Since the slope is the same anywhere on the solution line, this ratio** [fraction] **will _always_ be the same.**

The **number** that represents the slope of a line is the quantitative value for the "**steepness**" of the line. As the absolute value [the number with its sign ignored] of the slope number gets bigger, the line gets **steeper**. Think of walking **up** or **down** a hill. If the incline is gradual, the absolute value of the slope number would be small. As the hill gets steeper, the absolute value of the slope number gets larger. The **sign** of the slope [**positive or negative**] determines whether the line is going **up** or **down** [looking at the line left to right].

Note: You might be wondering why we are using **subscripts** [those numbers following x and y in the slope formula $m = \frac{y_2 - y_1}{x_2 - x_1}$]. They are to distinguish one **particular** x value from another and one **particular** y value from another.

   Example 1. Suppose I have determined that $(3,2)$, $(4,0)$, $(0,8)$ and $(5,-2)$ are four of the many solutions to the equation $2x + y = 8$. *How could I determine the slope of the solution line of this equation?*   [In mathematics terminology, we just say "What is the slope of the equation $2x + y = 8$?"]

As previously stated, we can assign **any** two solution points for (x_1, y_1) and (x_2, y_2) in the formula, $m = \dfrac{y_2 - y_1}{x_2 - x_1}$. We will **arbitrarily** assign two solution points $(3, 2)$ to (x_1, y_1) and $(5, -2)$ to (x_2, y_2) given above.

Using the formula: $\qquad m = \dfrac{y_2 - y_1}{x_2 - x_1}$

substituting for $\begin{matrix} 3 , 2 \\ (x_1, y_1) \end{matrix}$ and $\begin{matrix} 5, \ -2 \\ (x_2, y_2) \end{matrix}$: $\qquad m = \dfrac{-2 - 2}{5 - 3}$

we then **change subtraction to addition**: $\qquad m = \dfrac{-2 + (-2)}{5 + (-3)}$

performing the addition: $\qquad m = \dfrac{-4}{2}$

Reducing the fraction results in: $\qquad m = \dfrac{-2}{1}$ or -2

[is the *slope* of $2x + y = 8$]

We will obtain the same result using $\begin{matrix} (0 , 8) \\ (x_1, y_1) \end{matrix}$ and $\begin{matrix} (4, 0) \\ (x_2, y_2) \end{matrix}$

[These are two other solutions given for the equation $2x + y = 8$]

Using the formula: $\qquad m = \dfrac{y_2 - y_1}{x_2 - x_1}$

substituting for (x_1, y_1) and (x_2, y_2): $\qquad m = \dfrac{0 - 8}{4 - 0}$

we then **change subtraction to addition** in the numerator: $\qquad m = \dfrac{0 + (-8)}{4}$

performing the addition: $\qquad m = \dfrac{-8}{4}$

Reducing the fraction results in: $\qquad m = \dfrac{-2}{1}$ or -2

Notice that the resulting slopes are the same!

As an exercise, find other solutions to $2x + y = 8$ and use them to find the slope. Is the slope still -2 ?

The $y = mx + b$ Form of Linear Equations

As we have discussed previously, linear equations can **only** have variables that are raised to the **first power**. Therefore, an equation such as $2x + 3y = 12$ is **linear** and is understood to be $2x^1 + 3y^1 = 12$, although we will seldom see it written that way. The equation $2x^2 + 3y = 12$ is **not** linear since the x term in this case is to the second power. This equation would **NOT** have a *straight* solution line.

The linear equation $2x + 3y = 12$ is said to be in "**standard form**", when the x and y terms are on the **left side** of the equation **by themselves** and the constant [stand alone number] is on the **right side by itself**. However, we can *manipulate* [without changing its balance] the **standard form** into the

"$y = mx + b$" form for *any* linear equation [except the special linear equation in the form of $x = c$].
This new form has **many advantages** as we will see shortly.

Example 2.   Change $2x + 3y = 12$ from **standard form** to $y = mx + b$ form.

We will use the same strategy as **solving a literal equation**. We need the y term **by itself on the left side of the equation with a coefficient of** 1.

$$2x + 3y = \quad 12$$

We start by adding $-2x$ to both sides of the equation:

$$\underline{-2x \qquad\quad -2x}$$

Resulting in:

$$3y = -2x + 12$$

Notice that we cannot combine 12 and $-2x$ since they are not "like terms". Since the y term should have a coefficient of 1 ,we

Divide *every term* by 3 :

$$\frac{3y}{3} = \frac{-2x}{3} + \frac{12}{3}$$

Doing this creates the identity element, $\frac{3}{3}$, which equals 1 as the

y coefficient:

$$1y = \frac{-2x}{3} + 4$$

$$\left[\frac{-2x}{3} \Rightarrow \frac{-2 \cdot x}{3 \cdot 1} \Rightarrow \frac{-2}{3} \cdot \frac{x}{1} \Rightarrow -\frac{2}{3} \cdot x \Rightarrow -\frac{2}{3}x \right] \text{ Or just:} \qquad y = -\frac{2}{3}x + 4$$

The equation is now in the form:

$$y = mx + b$$

In this form, m [which is the value of the <u>coefficient</u> of x and, in this case, $-\frac{2}{3}$] **will always**

represent the *slope* **of the equation and** b [the value of the constant, 4 , in this case] will always represent the solution point on the $y - axis$ where the solution line touches [intersects] this axis. So, in this case, b is the y - **intercept** $(0,4)$. **Why is this?** As you remember from the previous section, in order to find the solution that is the y - **intercept** for an equation, **we must arbitrarily assign** 0 **to the** x **term** [Let $x = 0$] **in the equation.**

In this case, we will substitute 0 for x in this equation: $\qquad y = -\dfrac{2}{3}x + 4$

Then: $\quad y = -\dfrac{2}{3}(0) + 4$

Multiplying **any** number by 0 results in 0. So, regardless of the value of m: $\quad y = 0 + 4$

Or: $\quad y = 4$

The ordered pair solution in this case is $(0,4)$ which is the y-intercept.

Whenever an equation is in the form $y = mx + b$, "mx" will drop out when we assign x the value of 0. **Therefore,** $(0,b)$ **will always be the** y**-intercept when any linear equation is in this form!**

The $-\dfrac{2}{3}$ **slope** can be interpreted as follows: from **any solution point** on the solution line, to **any other solution point** on the solution line, the difference in y values $(y_2 - y_1)$ divided by the difference in x values $(x_2 - x_1)$ will reduce to $-\dfrac{2}{3}$ which is the same as $\dfrac{-2}{3}$ or $\dfrac{2}{-3}$ **[division by mixed signs is always negative!].**

Graphically, the slope, $\dfrac{y_2 - y_1}{x_2 - x_1}$, can be interpreted as $\dfrac{\text{the change in } y}{\text{the change in } x}$ or $\dfrac{\Delta y}{\Delta x}$ **[The Greek letter "delta"** (Δ) **means "change" in mathematics].** If we start at **any** point on the solution line and **go down** [the negative direction] **2 units** (-2) **[change in** y**]** and **from there go to the right** [the positive direction] **3 units** (3) **[change in** x**], we will be at another point on the solution line!**

If we wish to graph $y = \dfrac{-2}{3}x + 4$, we can start at the y-intercept that we know from the above discussion to be $(0,4)$. From there we use the slope, $\dfrac{-2}{3}$ to determine another point on the solution line **[See the graph below]**. If we start at $(0,4)$ and go down 2 units and to the right 3 units, we will arrive at a **second solution point** for the equation! As seen on the graph below, this second point is $(3,2)$. **From these two points we can draw the entire solution line!**

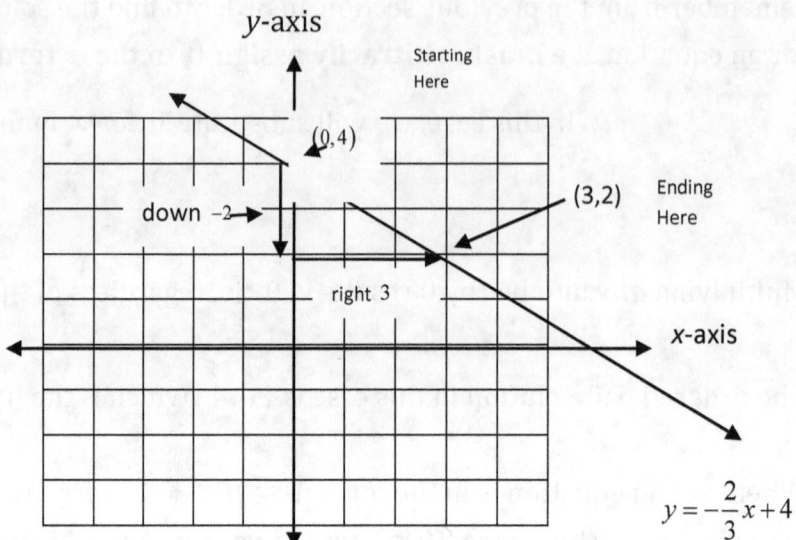

What do we know about every one of the points on the line that we created above? [They are all solutions for this equation!]

This is the "$y = mx + b$" form of the **equivalent equation** $2x + 3y = 12$ [standard form] that we started with in example 2. So, **to recap,**

> *Any* **linear equation, when put into its equivalent** $y = mx + b$ **form, will show us its slope** *(m)* **and y-intercept** *(0,b)*.

Example 3. **Applying** $y = mx + b$ **to model real life situations.**

Suppose a telecommunications company offers a plan for unlimited usage of a cell phone at $70 per month. The cost to purchase a phone is $40. How can we use this information to develop a linear equation to predict the monthly cost of the plan over various points in time?

The two variables defined in this situation are **cost** and **time**. There is a constant **rate of change** [synonymous with slope] in this situation since the cost is always $70 per month. The two variables can be represented by y, the cost variable which we wish to determine, and x, the amount of time. The cost of the phone is $40. Before any time has elapsed, the customer must purchase the phone for $40. Therefore, at the 0 point in time, the cost is 40. This is the definition of the y intercept or the $(0,b)$ in $y = mx + b$.

We have now modeled the problem as the linear equation: $y = 70x + 40$ where the m [rate of change or slope] is 70/1 month and b [y-intercept] is 40.

We can graph this equation showing all the "solutions". These solutions are all the **costs** [y variable] at the various **time intervals** [x variable]:

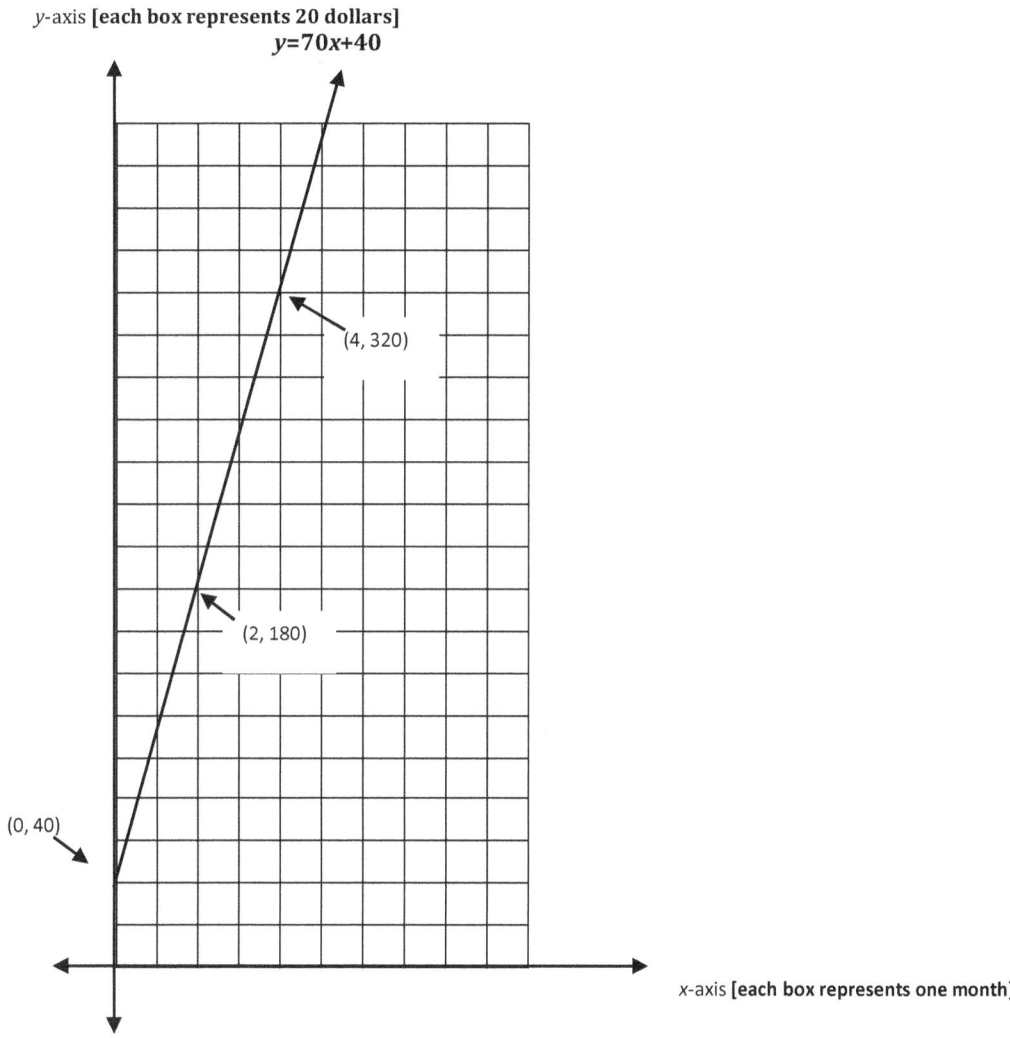

y-axis **[each box represents 20 dollars]**
$y=70x+40$

(4, 320)

(2, 180)

(0, 40)

x-axis **[each box represents one month]**

Every point on the solution line for $y = 70x + 40$ is solution for this equation. We can assign any value for x **[the time in months]** and find the corresponding cost at solution point on the line where these values intersect, which will be on the solution line for $y = 70x + 40$.

You might be wondering why the solution line does not extend lower than the y – intercept. This is becaue, in this model, it is non-sensical to think of time as a negative number. For example, it does not make sense to consider "negative 2 months".

Concept Homework

As an assessment of your understanding of the concepts set forth in this section, answer the following questions. If necessary, review the material to help you arrive at the correct conclusions. With "true or false" statements, if the statement is true, make up an example that supports the statement without using examples given in the material. If a statement is false, correct the wording to make it true. Then make up an example that supports the statement without using examples given in the material.

A. True or False

1. Linear equations can be expressed in **standard form** [variable terms on the left side - constant on the right side] or $y = mx + b$ [m is the slope and $(0,b)$ is the y –intercept] form and still be equivalent. The one exception are vertical lines [x=c] which have an undefined slope.

2. The $y = mx + b$ form of a linear equation is helpful because we will immediately know the slope of the equation as well as its y-intercept.

3. The y- intercept of an equation is that solution point on the solution line that is directly on the $x-axis$.

4. The ordered pair $(0,-1)$ would be the x - intercept of an equation.

5. Suppose we are given the equation $3x - 2y = -6$. We could find the slope by finding any two of its solution points.

6. Suppose we are given the equation $3x - 2y = -6$. We could find the slope by manipulating the equation without disrupting its balance so that $1y$ is alone on the left side of the equation.

7. Suppose we have the equation $y = 2x + 3$. The slope of the equation is such that if we start at any point on the solution line, go up two units, and from there go to the right 1 unit, we will arrive at another point on the solution line.

8. The slope of an equation can be determined by the formula $m = \dfrac{y_2 - y_1}{x_2 - x_1}$ where (x_1, y_1) and (x_2, y_2) are solutions to the equation.

9. The equations $m = \dfrac{y_2 - y_1}{x_2 - x_1}$ and $m = \dfrac{y_1 - y_2}{x_1 - x_2}$ will yield the same results since we can pick any solution point for (x_1, y_1) and any other solution point for (x_2, y_2).

Solutions:1. True; 2. True; 3. False; 4. False; 5. True; 6. True; 7. True; 8. True; 9. True

B. Using the graph paper below, make up **any** linear equation in **Standard Form** not already used as an example in this topic. Find two solution points for the equation and with those points, determine the slope of the equation using **both** formulas, $m = \dfrac{y_2 - y_1}{x_2 - x_1}$

and $m = \dfrac{y_1 - y_2}{x_1 - x_2}$. Re-write the equation in $y = mx + b$ form, and graph the equation using the y - intercept and slope.

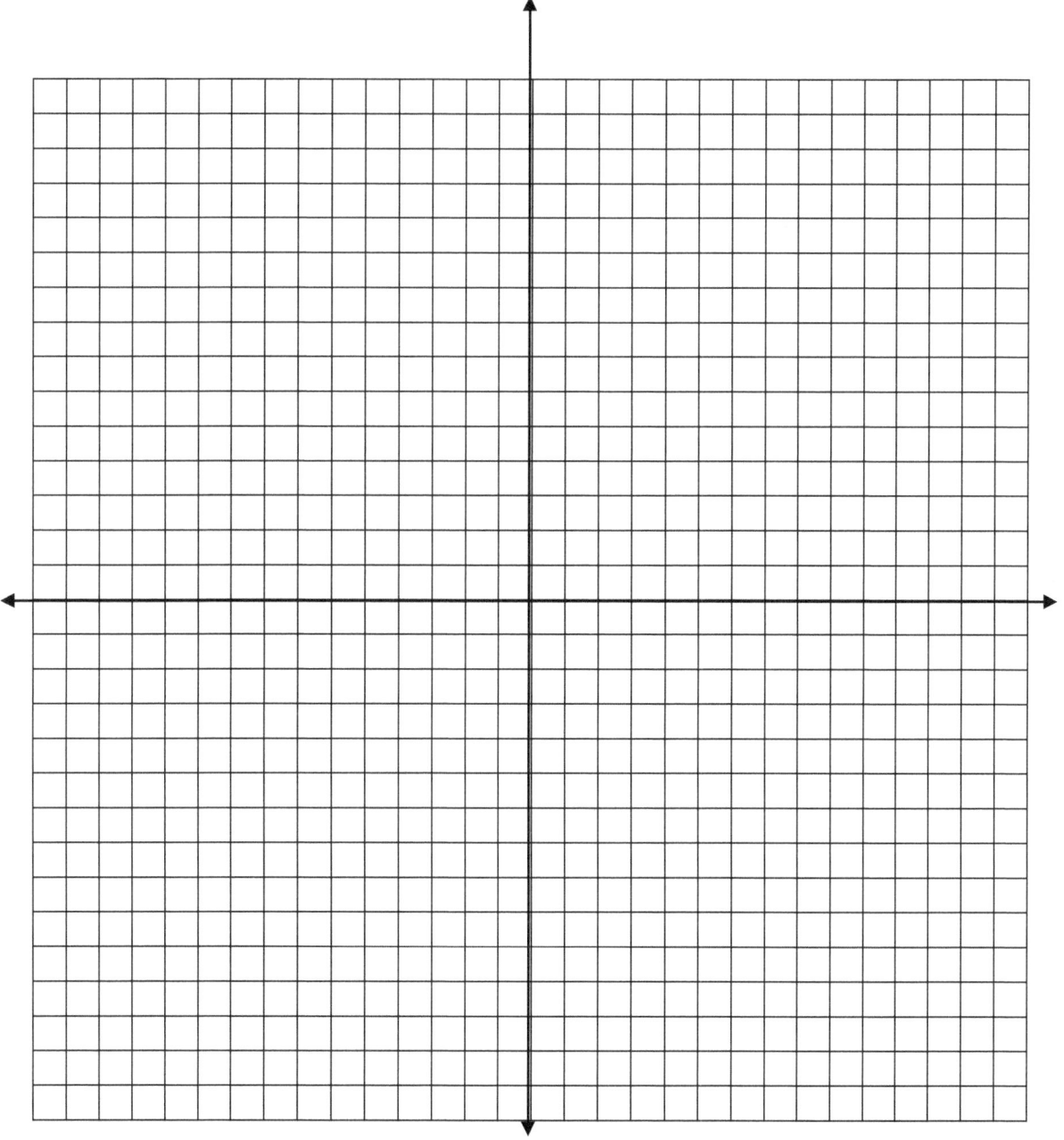

Exercises

1. Graph the lines represented by the following equations:

    a. $x + 2y = 2$ **b.** $x - 5y = 10$ **c.** $x + 2y = -3$ **d.** $2\,2x - 3y = 0$

2. Find the slopes of the lines going through the given solution points:

    a. $(2,1)\,\&\,(5,7)$ **b.** $(3,-5)\,\&\,(5,1)$ **c.** $(-4,-5)\,\&\,(11,7)$ **d.** $(-3,4)\,\&\,(-1,-2)$

3. Find the slopes of the following equations by converting the equations into $y = mx + b$ form:

    a. $2x - y = 0$ **b.** $2x + 7y = 0$ **c.** $7x + 8y = 10$ **d.** $2x - 3y = 6$

4. Using the same equations in **#3** above, find the slopes of each equation by finding two solutions for each equation and then using those solutions in the $m = \dfrac{y_2 - y_1}{x_2 - x_1}$ formula.

5. Find the slope of the equation $y = 4$ by finding two solutions for each equation and then using those solutions in the $m = \dfrac{y_2 - y_1}{x_2 - x_1}$ formula.

6. Find the slope of the equation $x = 4$ by finding two solutions for each equation and then using those solutions in the $m = \dfrac{y_2 - y_1}{x_2 - x_1}$ formula.

Solutions: 2a. $m = \dfrac{6}{3} = \dfrac{2}{1}$ or 2 ; 2b. $m = \dfrac{6}{2} = \dfrac{3}{1}$ or 3 ; 2c. $m = \dfrac{12}{15} = \dfrac{4}{5}$; 2d. $m = \dfrac{-6}{2} = \dfrac{-3}{1}$ or $\dfrac{3}{-1}$ or -3 ; 3a. $m = 2$ or $\dfrac{2}{1}$;

3b. $m = -\dfrac{2}{7} = \dfrac{-2}{7}$ or $\dfrac{2}{-7}$; 3c. $m = -\dfrac{7}{8} = \dfrac{-7}{8}$ or $\dfrac{7}{-8}$; 3d. $m = \dfrac{2}{3}$; 5. $m = 0$; 6. m is undefined .

Topic 13 - Deriving (Writing) Linear Equations from the Information Given

In this topic we will act as detectives and use the deductive reasoning process to determine the equation of a particular solution line given clues [information about the equation that will lead us to the equation itself]. In mathematics, this process is called **"deriving equations"**.

Suppose we are **given** the **slope** of a linear equation along with its y –intercept solution point. [recall that any y-intercept always has an x value of 0.] Could we determine what the equation is? Using the $y = mx + b$ form of linear equations, the task becomes simple!

Example 1a.

If the slope of a particular linear equation is $-\dfrac{3}{4}$, and the solution line of the equation crosses the $y-axis$ at $(0,-3)$, what is the equation for this solution line?

In our previous discussion, we have seen that *every* linear equation [except vertical lines: $x = c$] can be expressed in the form $y = mx + b$. The advantage of this form of a linear equation is that the "m" value [the coefficient of x in the equation], will *always* represent the slope of the equation line and $(0, b)$ [the constant in the equation] will *always* represent the y – intercept [graphically, where the solution line crosses the $y-axis$ at $(0,b)$].

Therefore, all we need to do is substitute the given slope, $-\dfrac{3}{4}$, for m, and -3 for b.

This is what happens: $y = mx + b$

but $m = -\dfrac{3}{4}$ and $b = -3$, so: $y = -\dfrac{3}{4}x - 3$

[solution]

As has been previously discussed, every equation in $y = mx + b$ form [as the equation above] can also be expressed in equivalent **standard form,** $[Ax + By = C]$, where A, B & C are usually integers, with the x and y terms on the left side of the equation with **coefficients** [represented by A and B], and the **constant** value on the right side of the equation [represented by C]. Let's see how we would approach this transformation:

Example 1b. **Change** $y = -\dfrac{3}{4}x - 3$ **to standard form** $[Ax + By = C]$. $\qquad y = -\dfrac{3}{4}x - 3$

Since standard form usually has integer coefficients,

we start with eliminating the fractional coefficient $\left[-\dfrac{3}{4}\right]$

of the x term without disrupting the balance of the
equation. This can be accomplished by multiplying **every**
term in the equation by the denominator, 4 , or its

equivalent, $\dfrac{4}{1}$, Whichever is **convenient**: $\quad (4)\,y = \left(\dfrac{\cancel{4}}{1}\right)\left(-\dfrac{3}{\cancel{4}}\right)x - (4)3$

The "canceling" results in: $\qquad 4y = \dfrac{-3}{1}x - 12$

We can now express the fraction $\dfrac{-3}{1}$ as a whole number: $\qquad 4y = -3x - 12$

Adding $3x$ to both sides of the equation: $\qquad \underline{3x \qquad\qquad 3x}$

Results in the equation being in **standard form**: $\quad 3x + 4y = \quad -12$

[solution]

Note: This is the same equation as $y - -\dfrac{3}{4}x - 3$ but in a different form!

Example 2.

Suppose we are given the **slope** and a **solution point** that is **not** the y – intercept
[remember that every y-intercept solution point *only* occurs when the x value of the ordered pair is 0].
How can we determine the equation? Here is an example of this case:

If the slope of a line is 2 and one solution for the equation is $(3, -1)$,

write the equation for this solution line.

Again, we are given " m ", the slope, **but notice that the x value of the given
ordered pair solution is *not* 0** . Therefore, it is not the y – intercept. How can we
determine the value of b ? Whenever we are asked to determine an equation, our

basic equation model should be: $\quad y = mx + b$

Since we are told that the slope is 2 , we can substitute this value for m: $\quad y = 2x + b$

[Don't be confused about the slope being a whole number. It can always be expressed

as a ratio or fraction $\dfrac{2}{1}$ if it is more convenient.]

We are given one solution to the equation $(3, -1)$. This means that

$x_1 = 3$, and $y_1 = -1$ is **a particular** solution that will satisfy the equation, as opposed to **any** solution (x, y). The use of ***subscripts*** with x and y denotes this idea of **one particular solution** of many such solutions. We can modify the model, $y = mx + b$, to represent one particular solution of x and y. By re-writing it as $y_1 = mx_1 + b$. Then the

equation looks like: $\quad y_1 = 2x_1 + b$

Since **we know** the value of x_1 and y_1 $[(3, -1)]$, we can plug them

into the equation: $\quad -1 = (2)(3) + b$

We are able to do this because we know that $(3, -1)$ is a solution to the equation. As can be seen, ***now there is now only one variable,*** "b" [the y-intercept] in the equation, so we can **find that value by isolating b.** First we simplify the right side: $\quad -1 = 6 + b$

Now, we add -6 to both sides of the equation: $\quad \underline{-6 \quad -6}$

Results in: $\quad -7 = 0 + b$

Eliminating the identity element for addition: $\quad -7 = b$

or its equivalent: $\quad b = -7$

Having found the value for b $[b = -7]$, we go back to the general

form of the equation: $\quad y = mx + b$

x and y now representing **all solutions** of the equation. Substituting 2 for m [that was given to us] and -7 for b [that we found]

we can write the equation : $\quad y = 2x - 7$

[solution]

An alternative way of solving this problem is to use the

"point/slope" formula: $\quad y - y_1 = m(x - x_1)$

This formula is derived directly from the definition

of slope $\left[m = \frac{y_2 - y_1}{x_2 - x_1} \right]$ and will also lead us go the same solution.

We start with by substituting the 2 for slope that we are given

for m in the formula: $\quad y - y_1 = 2(x - x_1)$

We now substitute the given solution point, $(3, -1)$, for

x_1 and y_1 respectively: $\quad y - (-1) = 2(x - 3)$

Notice that x and y in the equation remain as variables representing **all** the solutions to the equation. We now simplify the left side of the equation and distribute the right side: $\quad y + 1 = 2x - 6$

We now **isolate** y [get it alone on the left side of the equation] $y + 1 = 2x - 6$

by adding -1 to both sides: $\underline{-1 \qquad\; -1}$

Resulting in the same solution that we got using

the previous method: $y = 2x - 7$

[solution]

You will find that in some cases, the first method is more efficient to arrive at a solution and in other cases the second method makes the work easier. However, **both methods, if executed properly, will lead to the same correct solution.**

In order to understand the following examples, we must discuss some properties of **parallel** and **perpendicular** lines and their relationship to slope. As you might already know, parallel lines never intersect each other. This implies that **their slopes are the same**. Otherwise they would have to cross each other at some point and, therefore, would **not be parallel**.

For **example**, if solution Line A has a slope of 3 and is parallel to solution Line B, then Line B will also have a slope of 3. To summarize this idea:

Parallel lines have the same slope.

Perpendicular lines form right angles [90 degrees] at their intersection point. To see the relationship of the slopes of perpendicular lines, say Line A and Line B, suppose the slope of Line A is $\dfrac{a}{b}$. If line B is perpendicular to Line A, it's vertical change [Δy] and horizontal change [Δx] are interchanged. Also, The slope will have an opposite sign as the rotation of 90 degrees will reverse the slope's direction. Therefore, Line B will have a slope of $-\dfrac{b}{a}$. So,

Perpendicular lines will have slopes that are negative reciprocals.

For **example**, if solution Line A has a slope of $\dfrac{3}{4}$ and is perpendicular to solution Line B, then the slope of Line B will be $-\dfrac{4}{3}$.

Example 3. We will use the **theorem for parallel lines** to write the following equation:

If solution Line P has a solution point of $(3,-1)$ and is parallel to another solution Line Q whose slope is 2, what is the equation for Line P?

Since Lines P and Q are **parallel**, they have the **same slopes**. Therefore, Line P has a slope of 2. We can now use this slope, 2, and the given solution point $(3,-1)$ to write the equation. This is the same information given in **Example 2.** which resulted in the equation $y=2x-7$. Therefore Line P has an equation of: $y=2x-7$

[solution].

Example 4. We will use the **theorem for perpendicular lines** to write the equation for the following problem:

If a Line R has a solution point of $(3,-1)$ and is perpendicular to another Line S whose slope is $-\dfrac{1}{2}$, what is the equation for Line R?

Since Lines R and S are **perpendicular**, their slopes are **negative reciprocals**. Therefore, if Line S has a slope of $-\dfrac{1}{2}$, Line R's slope is **positive** and is the **reciprocal** of $\dfrac{1}{2}$ or $\dfrac{2}{1}$.

Therefore, its slope is $\dfrac{2}{1}$ or more simply, 2. We can now use this slope, 2, and the solution point $(3,-1)$ to write the equation just as was done in **Example 2** which will result in the equation for Line R being $y=2x-7$

[solution].

Example 5. Suppose we are just given **two solutions** [two ordered pairs] of an equation and **nothing else**. How do we determine the equation? Here is an example of this case:

Two solutions to an equation are $(1,-3)$ and $(-1,-1)$. Determine the equation for the line containing these points.

In this case we know **neither** the **slope nor** the y-intercept, so we will start by determining the **slope** of the equation. **Any** two solutions are enough to determine the slope of a solution line, since **the slope is constant** [consistent] anywhere along a straight line. We recall from our earlier discussion that the formula for slope is $m=\dfrac{y_2-y_1}{x_2-x_1}$, where (x_1, y_1) and (x_2, y_2) can be represented by **any** two solutions to the equation. We are **given two solutions,** so we assign the values $(1, -3)$ to (x_1, y_1) and $(-1, -1)$ to (x_2, y_2). **It is helpful** to use (x_1, y_1) and (x_2, y_2) as a **template** and take the time to write their corresponding values as shown below so that **we don't transpose** them in any way:

$$(1, \ -3) \quad \text{and} \quad (-1, -1)$$
$$(x_1, y_1) \qquad\qquad (x_2, y_2)$$

Now, when we plug them into our formula, it is easy to see which value goes with which variable!

$$m = \frac{y_2 - y_1}{x_2 - x_1}$$

Substituting $(1, -3)$ for (x_1, y_1) and $(-1, -1)$ for (x_2, y_2) results in:
$$m = \frac{-1 - (-3)}{-1 - 1}$$

We now can change the **subtraction operations to addition**:
$$m = \frac{-1 + 3}{-1 + (-1)}$$

Performing the addition operations, the resulting slope is:
$$m = \frac{2}{-2}, \text{ or } m = -1$$

The problem now is similar to that in **Example 2** except that we know the slope and **two** solution points, $(1, -3)$ and $(-1, -1)$. We can use **either** point to write the equation as we did in **Example 2**. *Either* point will result in the **same equation**. We will use $(1, -3)$ since it has already been labeled (x_1, y_1). We have the option either of the equation models $y_1 = mx_1 + b$ or, $y - y_1 = m(x - x_1)$.

Using the first equation model: $\quad y_1 = mx_1 + b$

We have already determined the slope $[m = -1]$, so we will plug it

in for m : $\quad y_1 = -1x_1 + b$

Now, plugging in for x_1 and y_1, the result is: $\quad -3 = -1(1) + b$

Simplifying: $\quad -3 = -1 + b$

To **isolate** b [the only variable left in the equation], we add 1 to both

sides of the equation:
$$\begin{array}{r} -3 = -1 + b \\ \underline{1 \quad\quad 1} \end{array}$$

Resulting in: $\quad -2 \ = \ 0 + b$

Or its equivalent: $b = -2$

Having found the value for b [$b = -2$], we go back to the: $y = mx + b$ model, keeping x and y **as variables**. Substituting -1 for m [that we found earlier] and -2 for b [that we just found] **we**

can write the equation: $y = -1x - 2$

or just: $y = -x - 2$

[solution]

The **second option**, $y - y_1 = m(x - x_1)$ will yield the same results: $y - y_1 = m(x - x_1)$

Substituting $m = -1$, $y_1 = -3$ and $x_1 = 1$: $y - (-3) = -1(x - 1)$

Simplifying and distributing results in: $y + 3 = -1x + 1$

Isolating y by adding -3 to both sides of the equation: $\underline{-3 \qquad -3}$

Results in: $y = -1x - 2$

Or just: $y = -x - 2$

[Solution]

Notice that the results are the same with **either** equation format. What would have happened if we used $(-1, -1)$ [the second solution point (x_2, y_2) given at the beginning of the example]? The equation model now becomes: $y_2 = mx_2 + b$ [since we are using the second solution point (x_2, y_2)]

We have already determined the slope [$m = -1$] , so we will plug it

in for m : $y_2 = -1x_2 + b$

Now, plugging in for $x_2 = -1$ and $y_2 = -1$, the result is: $-1 = -1(-1) + b$

Simplifying: $-1 = 1 + b$

Adding -1 to both sides of the equation: $\underline{-1 \quad -1}$

Results in: $-2 = 0 + b$

Or its equivalent: $b = -2$

Having a value for b [$b = -2$], we go back to the general $y = mx + b$ model, keeping x and y **as variables**. Substituting -1 for m [that we found earlier] and -2 for b [that we just found]

We can write the equation: $y = -1x - 2$

Or just: $y = -x - 2$

[solution]

As can be seen. the resulting equation is the same for either $(-1,3)$, or $(-1,-1)$. This is because both $(-1,3)$, or $(-1,-1)$ are solution points for this equation!

Note: If you use $(-1,-1)$ in the $y - y_2 = m(x - x_2)$ equation model along with the slope, $[m = -1]$, the same equation, $y = -x - 2$, will also be the result.

To recap: **Whenever we are asked to derive an equation, we start with the general form, $y = mx + b$, and then go about finding the value of m and b. We can find these values if we are given the following information:**

- **Either we are given: The slope (m) and the y-intercept (b)**
  We just substitute these values into the equation $y = mx + b$ [see example 1a].
- **Or we are given: The slope (m) and a solution point that is *not* the y-intercept**
  (a) We use the slope *(m)* and the solution point (x_1, y_1) in the equation
  $y_1 = mx_1 + b$ to find b, or as an alternative,
  (b) in the point/slope formula, $y - y_1 = m(x - x_1)$ [see example 2].
- **Or we are given: Two solution points for the equation.**
  (a) We find the slope **(m)** with the two solution points using them as (x_1, y_1)
and
  (x_2, y_2) in the formula $m = \dfrac{(y_2 - y_1)}{(x_2 - x_1)}$ [see example 5].

  (b) We use the slope *(m)* and **either** solution point (x_1, y_1) or (x_2, y_2) in the
  equation $y_1 = mx_1 + b$ to find b, or as an alternative, in $y - y_1 = m(x - x_1)$

We also know that **parallel lines** have the **same slope** and **perpendicular lines** have slopes that are **negative reciprocals**. [See examples 3 & 4.]

Concept Homework

As an assessment of your understanding of the concepts set forth in this section, answer the following questions. If necessary, review the material to help you arrive at the correct conclusions. With "true or false" statements, if the statement is true, make up an example that supports the statement without using examples given in the material. If a statement is false, correct the wording to make it true. Then make up an example that supports the statement without using examples given in the material.

A. <u>True or False</u>

1. ***Every*** linear equation [except those for vertical lines] in the form $Ax + By = C$ can be expressed in the $y = mx + b$ form in such a way that they will still be equivalent equations.

2. A linear equation in standard form, $Ax + By = C$ is helpful because we can immediately determine the slope of the equation as well as its y-intercept.

3. The y-intercept of an equation is that solution point that has 0 for the y-value of its ordered pair.

4. The ordered pair $(-1, 0)$ could be the x-intercept of some equation.

5. The linear equation $y = 4x$ is a linear equation where the y-intercept is also the x-intercept. **[If you graph this equation, the answer will become clear.]**

6. The linear equation $x = -3$ has no y-intercept.

7. Suppose we are given the equation $3x - 2y = -6$. We could find the slope by finding two solutions for this equation.

8. Suppose we are given the equation $3x - 2y = -6$. We could find the slope by manipulating the equation into $y = mx + b$ form.

9. Suppose we have the equation $y = 2x + 3$. The slope of the equation is such that **if we start at $(0, 3)$ on the solution line**, move up two units, and from there, move 1 unit to the right, we will arrive at **another point** on the solution line.

10. We cannot write an equation unless one of the two solution points given to us is the y-intercept.

11. If we know the slope of a line as well as the solution point whose x value is 0, we can easily write the equation for the solution line without making any calculations.

Solutions: 1. True; 2. False; 3. False; 4. True; 5. True; 6. True; 7. True; 8. True; 9. True; 10. False; 11. True.

Exercises

Write the equations of the following lines given the following information. Write the equations in

both $y = mx + b$ and standard form::

    a. Slope of 5; solution point $(0,6)$.

    b. Slope of $-\dfrac{3}{5}$; solution point $(0,0)$

    c. $m = 4$; solution point $(1,3)$.

    d. $m = -\dfrac{1}{5}$; solution point $(4,0)$

    e. Parallel to the solution line $y = -8x + 3$ and has a solution point $(-11,-12)$

    f. Perpendicular to the solution line $y = -\dfrac{2}{3}x + 5$ and has a solution point $(5,-6)$

    g. Solution points $(3,2)$ and $(5,6)$

    h. Solution points $(2,3)$ and $(-1,-1)$

Solutions:

a. $y = 5x + 6$; $-5x + y = 6$; **b.** $y = -\dfrac{3}{5}x$; $3x + 5y = 0$; **c.** $y = 4x - 1$; $-4x + y = -1$ or $4x - y = 1$; **d.** $y = -\dfrac{1}{5}x + \dfrac{4}{5}$; $x + 5y = 4$;

e. $y = -8x - 100$; $8x + y = -100$; **f.** $y = \dfrac{3}{2}x - \dfrac{27}{2}$; $-3x + 2y = -27$ or $3x - 2y = 27$; **g.** $y = 2x - 4$; $-2x + y = -4$ or $2x - y = 4$;

h. $y = \dfrac{4}{3}x + \dfrac{1}{3}$; $-4x + 3y = 1$ or $4x - 3y = -1$

Topic 14 - Systems of Equations

If we graph two different solution lines on a coordinate system they could only intersect [meet] each other at **one and only one point**. If you experiment with this idea of any intersecting straight lines you will see that this is true. However, if the lines were parallel to each other, they would **never** intersect. This happens when the **slopes of such lines are equal.**

We have learned that the graph of **every straight line** on a coordinate system represents **the solution set to a particular linear equation.** We can derive [determine] such an equation using the techniques previously discussed in Topic 13 [**finding the slope and y-intercept of the line and plugging it in to the model** $y = mx + b$].

The graph of a linear equation [**which is always a straight line**] is really a series of ordered pairs [x and y values] that all "satisfy" the equation [**when plugged into the equation, they make it true**]. If we have two different lines on the same coordinate system, the point at which they will intersect [meet] has **special significance.** Two such lines, as well as the equations that they represent are called a **"system of equations".**

That particular point where two lines intersect satisfies [makes true] <u>both</u> **linear equations represented by these lines. For any two non-equivalent linear equations there can be at most,** *only one such point.*

In other words, **the x and y values represented by this intersection is the unique point that will make <u>both</u> equations true.** We can determine this point by **"solving" the system of equations.**

Let's look at an example to make this idea more concrete. One way to solve a system of equations is by:

<u>The Addition Method</u>

Example 1.

What **unique** [only one] ordered pair will make both of the following **system of equations** true?   [We are looking for an x and y value that will "satisfy" <u>both</u> equations!]

First equation:	$x + y = 7$
Second equation:	$x - y = 3$

How can we determine which *x* and *y* values will make both of these equations true? **We must find a way to manipulate both equations in such a way that we <u>eliminate a variable</u>.** What would happen if we **added the terms on the left side of the equations together as well as those on the right side of the equations?** Would the result still be a valid equation? **YES,** because we are adding equal quantities [first equation's left side equals its right side] to equal quantities [second equation's left side equals its right side]. Therefore, the results must also be equal!

Let's add the two equations [We will put in "1" coefficients to make the addition clearer.]:

$$1x + 1y = 7$$
$$1x - 1y = 3$$

Adding the equations results in: $2x + 0 = 10$

The addition caused the *y*-variable to be eliminated!

Eliminating the identity element for addition, 0, results in: $2x = 10$

Dividing both sides of the equation by 2: $\dfrac{2x}{2} = \dfrac{10}{2}$

Results in: $1x = 5$

Or just: $x = 5$

Our first objective is accomplished since we know that the *x* value must be 5.

Now, we need to determine the corresponding y value when $x = 5$. This will determine the ordered pair that will solve **both** equations. We can accomplish this by assigning 5 to the x variable in ***either*** of the equations. [The result will be the same no matter which equation we decide to use!]

Suppose we let $x = 5$ in the first equation: $1x + 1y = 7$

Plugging in $x = 5$: $(1)(5) + 1y = 7$

Simplifying: $5 + 1y = 7$

Adding -5 to both sides: $\underline{-5 \qquad -5}$

Results in: $0 + 1y = 2$

Or more simply: $y = 2$

With these values for *x* and *y*, we have found the **unique ordered pair that satisfies both equations**:

$$x = 5 \text{ and } y = 2, \text{ or the ordered pair:} \quad (5, 2) \quad \text{[solution]}$$

We can verify that this ordered pair **indeed satisfies both equations** by plugging in the values for *x* and *y* into **both** of the equations to see if they make them true:

First equation: $x + y = 7$

plugging in 5 for x and 2 for y: $5 + 2 = 7$ [true]

Second equation: $x - y = 3$

plugging in 5 for x and 2 for y: $5 - 2 = 3$ [true]

So, we've verified that $(5,2)$ **does indeed satisfy both equations!**

Another method for solving these types of equations is called

The Substitution Method.

Using the same system of equations: (1) $1x + 1y = 7$

(2) $1x - 1y = 3$

We **isolate** [get it alone on one side of the equation] one variable in **either** equation. We will isolate x in equation (1) $1x + 1y = 7$

$$\underline{-1y \quad -1y}$$

Resulting in: $1x = -1y + 7$

Now the system looks like (1) $1x = -1y + 7$

(2) $1x - 1y = 3$

From equation (1), we now substitute the equivalent value for $1x$ (1) $1x = \boxed{-1y + 7}$

which is $-1y + 7$, for $1x$ in equation (2).   (2) $1x - 1y = 3$

which results in: $-1y + 7 - 1y = 3$

Combining like terms: $-2y + 7 = 3$

Adding -7 to both sides of the equation: $\underline{-7 \quad -7}$

Results in: $-2y = -4$

Dividing both sides of the equations by -2: $\dfrac{-2}{-2}y = \dfrac{-4}{-2}$

Results in: $1y = 2$

Or just: $y = 2$

This is the same result that we got using the **addition method** previously!

Now, we assign 2 to the y variable in ***either*** of the original equations.

Using equation (2): $1x - 1y = 3$

The equation becomes: $1x - 1(2) = 3$

or: $1x - 2 = 3$

Adding 2 to both sides of the equation: $\underline{2 \quad 2}$

We get the same value for x as we did with the **addition method**: $1x = 5$

or just: $x = 5$

Our solution for the system is the same as with the addition method: $(5,2)$

Let's graph both of these equation lines on the coordinate system below:

We'll do this by finding the x and y intercepts for each equation:

Graphing: $x + y = 7$	$x + y = 7$
Let $x = 0$: $0 + y = 7$	Let $y = 0$: $x + 0 = 7$
Then: $y = 7$	Then: $x = 7$
$(0,7)$ (_y - intercept_)	$(7,0)$ (_x - intercept_)

Graphing: $x - y = 3$	$x - y = 3$
Let $x = 0$: $0 + (-1)y = 3$	Let $y = 0$: $x + -1(0) = 3$
dividing both sides by $^{-}1$: $\dfrac{-1y}{-1} = \dfrac{3}{-1}$	simplifying: $x + 0 = 3$
results in: $y = -3$	results in: $x = 3$
$(0,-3)$ (_y - intercept_)	$(3,0)$ (_x - intercept_)

We will graph these two lines, using their x and y intercepts:

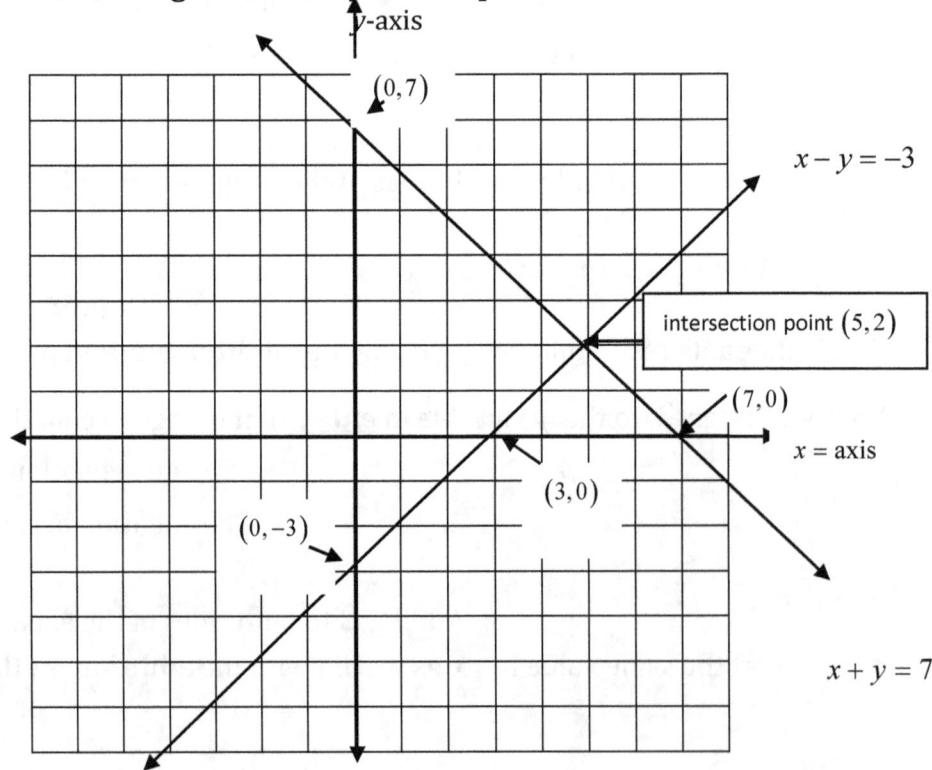

We see that the lines intersect at $(5,2)$. This is the same ordered pair solution that we found earlier using *algebraic approaches* on the preceding pages!

Here is an important question to consider: Could there be additional points where these two solution lines intersect? When we look at the graph, we see that the answer is **NO, since the equation lines are straight, there can only by one intersection point and one ordered pair that will solve both of these equations.**

Lets look at another example of solving a **"system of two equations"**.

Example 2. Find the unique [only one] solution for the following **system of equations**:

$$3x - y = -2 \qquad \text{adding a coefficient} \Rightarrow \qquad 3x - 1y = -2$$
$$x + 5y = 10 \qquad \text{adding a coefficient} \Rightarrow \qquad 1x + 5y = 10$$

Using the **addition method**, the problem with this system is that adding the two equations together will **not** result in either x or y dropping out as happened in the first example.

If we add these equations together, we will **not eliminate either x or y** which is necessary. We therefore have to find a number, that when multiplied through either equation will create *opposites* for either the x or y terms. When added, **opposites will become 0!**

[Note: Sometimes it is necessary to multiply both equations by different numbers to achieve our goal of eliminating a variable. In the subsequent Example 4, we will demonstrate this.]

We can choose <u>any multiplier that we wish</u>. **As long as we multiply it through the entire equation, the "new" equation will still be true.** [This is the *multiplication property of equations*!]

In this case, we can ***arbitrarily*** [we can choose any number we wish] choose -3 as a multiplier for the **second equation** because it creates the opposite of $3x$ in the **first equation**:

$$\text{first equation:} \qquad\qquad\qquad\qquad\qquad 3x - 1y = -2$$
$$\text{second equation:} \quad -3(1x + 5y = 10) \quad \Rightarrow \quad \underline{-3x + -15y = -30}$$

Now when we add the equations, the **x terms are eliminated:** $\qquad 0 \;-16y = -32$

eliminating the 0 and dividing both sides of the equation by -16: $\qquad \dfrac{-16y}{-16} = \dfrac{-32}{-16}$

Results in: $\qquad 1y = 2$

or just: $\qquad y = 2$

Now that we have found the solution for y, **we use it in _either equation_ to find the corresponding x value. We will then know the ordered pair solution that makes both equations true**. This time we will use the **second** equation [we could have used either]: $\quad 1x + 5y = 10$

we **assign y the value of** 2 [what we found above] to find the solution for *x*: $\quad 1x + 5(2) = 10$

$$\begin{aligned} \text{performing the multiplication:} \quad & 1x+10 = 10 \\ \text{Adding } -10 \text{ to both sides:} \quad & \underline{-10 \quad -10} \\ \text{Results in:} \quad & 1x = 0 \\ \text{Or just:} \quad & x = 0. \end{aligned}$$

Therefore, the **unique** solution that satisfies **both** of these equations is: $x = 0$ and $y = 2$

Or, as an ordered pair: $\left(0,2\right)$ [solution]

Let's graph these two equations, $\quad 3x - 1y = -2 \quad$ and $\quad 1x + 5y = 10$

Converting the equations to $y = mx + b$ form: $\underline{-3x -3x} \qquad \underline{-1x -1x}$

On the right sides of the equations, the terms cannot be combined since they are not "like": $\quad -y = -3x - 2 \qquad\qquad 5y = -1x + 10$

Making the coefficients of the y term 1and

Doing the necessary divisions: $\quad \dfrac{-1y}{-1} = \dfrac{-3x}{-1} + \dfrac{-2}{-1} \qquad \dfrac{5y}{5} = \dfrac{-1x}{5} + \dfrac{10}{5}$

Resulting in: $\quad y = \dfrac{-3}{-1}x + 2 \qquad\qquad y = \dfrac{-1}{5}x + 2$

We now know the slopes and y-intercepts of each equation, making the graphing easy:

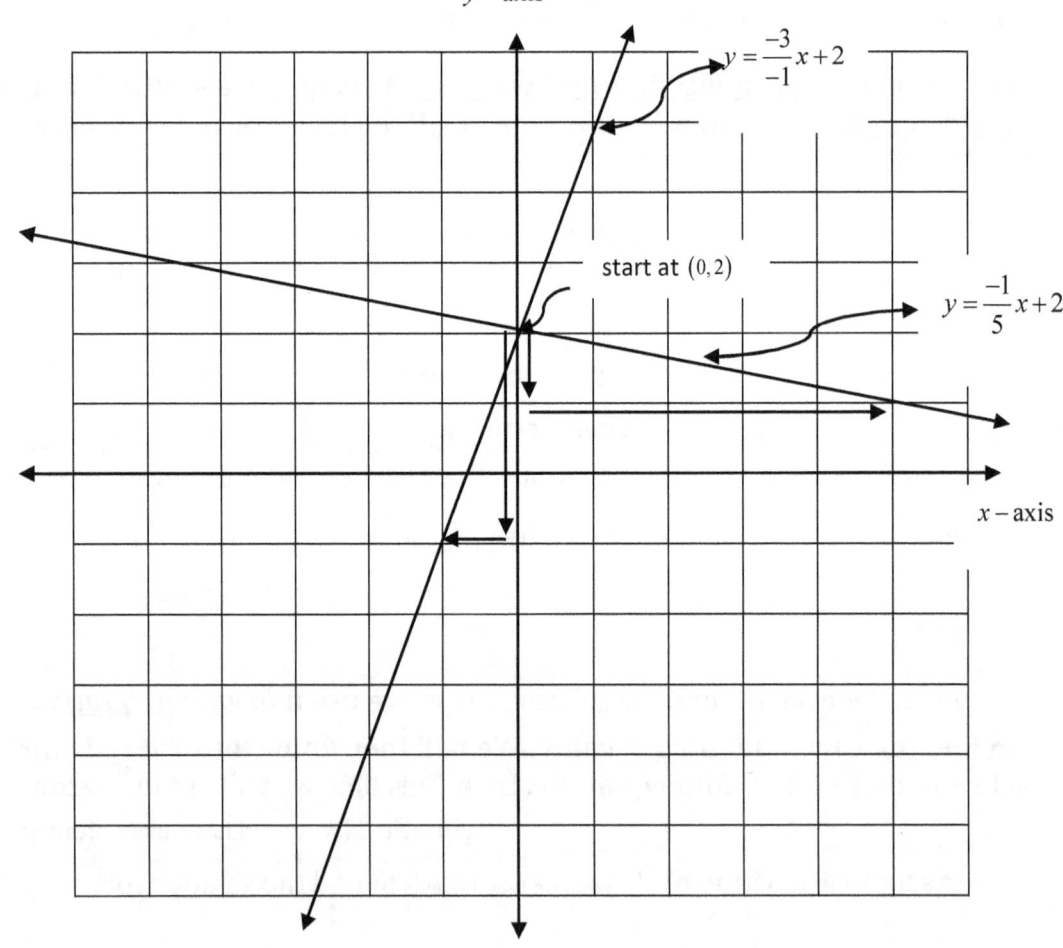

As can be seen, the intersection point on the graph of the two equations is $(0,2)$. This is the same ordered pair that we determined algebraically on the preceding page!

Now let's look at the case where lines are parallel. Since parallel lines never meet, there can be **no intersection point.** Therefore, if we have a system of equations whose solution lines are parallel to each other, there will be **no solution** to the system. Here is an example:

Example 3. Find the solution to the system of equations: **[The following equations have solution lines that are parallel! This can be seen from the graphs on the following page. Notice from the boxes below, the slopes are the same!]**

$$x + y = 5 \quad \Rightarrow \quad 1x + 1y = 5$$
$$x + y = 2 \quad \Rightarrow \quad 1x + 1y = 2$$

[Notice that the x and y terms are *identical*. Only the constants on the right side of the equations are different!]

In order to get the x variable to drop out we could multiply the second equation by -1:

$$1x + 1y = 5 \quad \Rightarrow \text{[stays as is]} \quad 1x + 1y = 5$$
$$-1(1x + 1y = 2) \quad \Rightarrow \quad -1x - 1y = -2$$

But, when we add these equations together, **both x and y become 0:** $\quad 0 + 0 = 3$
$$0 \neq 3$$

This shows us that there is no solution to the system!

The only way for this to happen is if the **lines were parallel,** which would mean that **their slopes are equal.** These equations are graphed on the coordinate system on the following page and we see that they are indeed parallel. Let's put each equation in $y = mx + b$ form:

First Equation: $\quad 1x + 1y = \quad 5$ **[standard form]**
adding $-1x$ to both sides: $\quad -1x \qquad -1x$
results in: $\quad y = -1x + 5 \quad$ or $\quad y = \dfrac{-1}{1}x + 5 \quad$ **[$y = mx + b$ form]**

Second Equation: $\quad 1x + 1y = \quad 2$ **[standard form]**
adding $-1x$ to both sides: $\quad -1x \qquad -1x$
results in: $\quad y = -1x + 2 \quad$ or $\quad y = \dfrac{-1}{1}x + 2 \quad$ **[$y = mx + b$ form]**

Notice that their *slopes* (*m*) are equal [both $\frac{-1}{1}$]. This is not a coincidence! All parallel lines have equal slopes!

When you look at the lines below, you see that their *steepness* is constant and the same! The only difference between the graphs of the equations are their y-intercepts which determines their "vertical" locations on the graph.

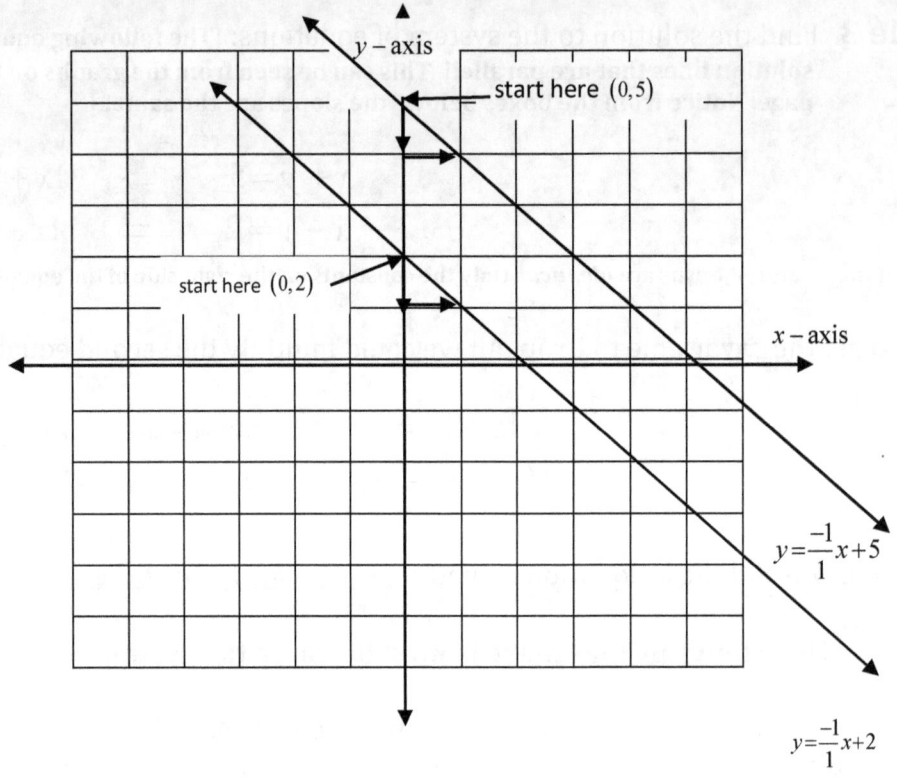

$x + y = 5$ [equivalent to $y = \frac{-1}{1}x + 5$] and $x + y = 2$ [equivalent to $y = \frac{-1}{1}x + 2$] **are parallel** since their **slopes are equal!**

Is there a **practical application** for this concept of **systems of equations**? THERE ARE MANY!

Example 4: Application

> Julia and Marcia were shopping at the same grocery store and bought identical containers of chili and salsa to make a dip.
> Marcia bought 2 containers of chili and 4 containers of salsa for a total of $12.98.
> Julia bought 3 containers of chili and 2 containers of salsa for a total of $10.07.
> How much was the cost for one container of chili and one container of salsa?

This information represents **two linear equations** that we can use as a **system** to solve this problem!

First, we **define the two variables** contained in the problem:
$$c = \text{ cost of a container of chili}$$
$$s = \text{ cost of a container of salsa}$$

We can write Julia's purchase mathematically as: $\quad 2c + 4s = 12.98$
We can write Marcia's purchase mathematically as: $\quad 3c + 2s = 10.07$

We can now solve this system! We will use **the addition method** to eliminate a variable when adding these equations together. We will have to multiply **both equations** by some arbitrary [we choose whichever we want] multiplier to create opposites for one of the variables. **It does not matter which variable $\left[c \text{ or } s\right]$ we choose to eliminate!**

We will arbitrarily choose to concentrate on the "c" variable. We notice that $2c$ and $3c$ have a common multiple in 6, so we can transform the first equation so that the $2c$ becomes $6c$, and then we transform the second equation so that the $3c$ becomes $-6c$. By doing this, we will have created **"opposite terms"** that will become "0" when added. We can accomplish this by **multiplying _every term_** in each equation by the appropriate multiplier as shown below:

multiplying by 3: $\quad 3\left(2c + 4s = 12.98\right) \quad \Rightarrow \quad 6c + 12s = 38.94$

multiplying by -2: $\quad -2\left(3c + 2s = 10.07\right) \quad \Rightarrow \quad \underline{-6c - 4s = -20.14}$

We now have **opposite terms** $\left[6c \text{ and } -6c\right]$,
so when we add the equations, the results are: $\qquad 0 + 8s = 18.80$

dividing both sides by 8: $\qquad \dfrac{8s}{8} = \dfrac{18.80}{8}$

results in: $\qquad 1s = 2.35$

So we have determined that Julia and Marcia paid $2.35 **for a container of salsa.**

Knowing that a container of salsa (s) costs 2.35, we can substitute that amount for the "s" variable in **either** equation to find out the cost of a container of chili (c):

Let $s = 2.35$ in the equation	$2c + 4s = 12.98$
Results in:	$2c + 4(2.35) = 12.98$
Doing the multiplication:	$2c + 9.40 = 12.98$
Adding -9.40 to both sides of the equation:	$\underline{-9.40 \quad -9.40}$
Results in:	$2c + 0 = 3.58$
Eliminating the 0 and dividing both sides by 2:	$\dfrac{2c}{2} = \dfrac{3.58}{2}$
Gives us:	$1c = 1.79$

We have determined that Julia and Marcia paid $\$1.79$ **for a container of chili.**

To insure that we did not make an error in our calculations, we should take these values for c and s and plug them back into **both** equations to make sure that they are indeed a solution for each:

$c = 1.79$ and $s = 2.35$ into:	$2c + 4s = 12.98$
plugging in:	$2(1.79) + 4(2.35) = 12.98$
multiplying results in:	$3.58 + 9.40 = 12.98$
adding results in:	$12.98 = 12.98$

[$c = 1.79$ and $s = 2.35$ makes the equation true!]

It is **not sufficient** to show that these values satisfy one equation. They **must satisfy both**, so it is necessary to **plug these values into the other equation also**:

$c = 1.79$ and $s = 2.35$ into:	$3c + 2s = 10.07$
Plugging in:	$3(1.79) + 2(2.35) = 10.07$
Multiplying results in:	$5.37 + 4.70 = 10.07$
Adding results in:	$10.07 = 10.07$

[$c = 1.79$ and $s = 2.35$ also makes this equation true!]

Concept Homework

As an assessment of your understanding of the concepts set forth in this section, answer the following questions. If necessary, review the material to help you arrive at the correct conclusions. With "true or false" statements, if the statement is true, make up an example that supports the statement without using examples given in the material. If a statement is false, correct the wording to make it true. Then make up an example that supports the statement without using examples given in the material.

True or False:

    a. If we have any two unique linear equations whose slopes are different, there will be two unique ordered pair solutions that will make both of these equations true.

    b. The solution lines of any two linear equations will always intersect at some point.

    c. We can *always* manipulate a system of equations, whose graphs are <u>not parallel</u>, so that one of the variables can be eliminated.

    d. On a graph, the intersection point of two solution lines represents the unique values for x and y that will satisfy [make true] both equations.

    e. If we add two linear equations in a system together and both variables "drop out" [take on zero values], then there will still be one solution to the system.

    f. The solution for a system of equations that have the same slope is always $(0,0)$.

    g. We can find the solution to a system of equations [if there is one] either algebraically or graphically.

    h. Although this idea of systems of equations is interesting to mathematicians, it has no practical usefulness in the real world.

    i. Using the addition or substitution method for solving a system of equations will always result in the same solutions.

    j. If we graph a system of equations that have no solution, we will find that their solution lines are perpendicular.

Solutions: a. False; b. False; c. True; d. True; e. False; f. False; g. True; h False; i. True; j. False.

Exercises

1. Use the addition method, the substitution method, and graphic method to find the solution to the following equations: **[Graph paper can be obtained free online.]**

a. $\begin{aligned} x + y &= 6 \\ 3x - y &= 2 \end{aligned}$ **b.** $\begin{aligned} x - y &= 5 \\ 3x + 2y &= 5 \end{aligned}$ **c.** $\begin{aligned} x - 2y &= 0 \\ 2x - y &= 6 \end{aligned}$ **d.** $\begin{aligned} 2x + 6y &= 11 \\ x + 3y &= 3 \end{aligned}$

2. Use the addition method to solve the following equation: $\begin{aligned} 3x + 2y &= 12 \\ 5x - 3y &= 1 \end{aligned}$

3. Use the substitution method to solve the following equation: $\begin{aligned} 5x + y &= 8 \\ 3x - 2y &= 10 \end{aligned}$

4. If 5 pounds of almonds and 4 pounds of walnuts cost $14, while 8 pounds of almonds and 6 pounds of walnuts cost $21.80, what is the price of each per pound?

Solutions: 1a. $(2,4)$; 1b. $(3,-2)$; 1c. $(4,2)$; 1d. No solution; parallel lines; 2. $(2,3)$; 3. $(2,-2)$; 4. almonds: $1.60 & walnuts: $1.50.

Topic 15 – Further Applications for Systems of Equations

As we have seen earlier, two linear equations [that have two different solution lines that are not parallel.] each have an infinite number of solutions represented graphically by two solution lines. However, both equations taken together as a "system", have only one solution: a **unique** [only one] ordered pair $(x-value,\ y-value)$ that solves **both** of the equations. Graphically, this value is located at the intersection of the solution lines. It is the only solution point on the graph that lies on **both** solution lines.

How can we use this mathematical concept to solve "everyday" problems?

Many situations present themselves where two linear equations can be developed modeling two different kinds of information **using the same variables.** After establishing these systems of two equations, we will solve them **algebraically** rather that **graphically**.

Example 1.

We have change in our pocket that consists of only nickels and quarters. We know that there are 22 coins and they have a total value of $3.50. How many nickels and how many quarters there are?

In examining this problem, we should always start by considering what we are asked to find. This helps us decide how to define the variables. In this case, we want to know **how many** nickels and **how many** quarters there are. Therefore, the variables should be: the **number** of nickels (n) and **number** of quarters (q). Also, the problem gives us **two different kinds of information** that can be **expressed in terms of these variables**:

> **(1) the number of coins**
> **(2) the total dollar value of the coins**

Can we develop two separate linear equations from this information? **YES!**

Let's consider **(1): the number of coins.**

We are told that there are a total of 22 coins. We don't at this point know how many quarters and nickels there are [our objective is to find out!] but we do know that if we add the number of coins together, they will total 22. This would take the form of **our first linear equation**:

$$(1) \quad n + q = 22.$$

138

Now let's consider **(2): the total dollar value of the coins.**

We know that each quarter has a value of 25 cents [$0.25], and each nickel has a value of 5 cents [$0.05] If we multiply .25 by q [the variable for <u>number</u> of quarters], and we multiply 0.05 by n [the variable for the <u>number</u> of nickels] and add them together, the amount should total $3.50 [information given to us in the problem]. This would take the form of **a second linear equation:**

$$(2) \quad .05n + .25q = 3.50.$$

We now have a *system* of two linear equations using the *same* variables. Therefore, there is only one value for n and q that will make both equations true. When we find those values, we will have solved this problem:

$$(1) \qquad n + q = 22$$
$$(2) \quad .05n + .25q = 3.50$$

Decimal points confuse the issue, so we will multiply through equation (2) by 100:

$$100(.05n + .25q = 3.50) \Rightarrow 5n + 25q = 350$$

[This is the multiplication property of equations!]

Inserting the implied coefficients, the first equation looks like: (1) $\quad 1n + 1q = 22$
The transformed second equation without decimals is: (2) $5n + 25q = 350$

We see that **arbitrarily** multiplying equation (1) by -5 will
be *convenient*: $\qquad$ (1) $-5(1n + 1q = 22) \Rightarrow -5n - 5q = -110$
$\qquad$ (2) [leaving the equation as is] $\Rightarrow 5n + 25q = \quad 350$

This leads to the n variable "dropping out" when we add the
equations together : $\qquad\qquad 20q = \quad 240$

Dividing both sides of the equation by 20 : $\qquad \dfrac{20q}{20} = \dfrac{240}{20}$

Results in: $\qquad 1q = 12$

Or just: $\qquad q = 12$

[The number of quarters]

So now we know that there has to be **12 quarters**.

{ 138 }

We use this information in equation (1) **[we could also use equation (2) for this step]**

To find the value for n, we let $q=12$ in $n + q = 22$.

$$\text{Substituting for } \boldsymbol{q}: \quad n + 12 = 22$$
$$\text{Adding } -12 \text{ to both sides of the equation:} \quad \frac{-12 \quad -12}{}$$
$$\text{results in:} \quad n \quad = 10$$

[the number of nickels]

Therefore, there must be 12 nickels!

We have answered the question that was asked in the problem:

We have 12 quarters and 10 nickels in our pocket.

Now to **verify that are solutions are correct**, we see if they satisfy **both** equations that we have created:

(1) Is there 22 coins altogether **[according to the given information, this must be so.]**? **Yes**! [12 + 10 = 22]

(2) Also, the **value** of these coins must equal $3.50. **[according to the given information, this must be so.]**
   Does it? **Yes!**

The **value** of 12 quarters is **$3.00** and the **value** of 10 nickels is **$0.50** for a total of **$3.50**.

Example 2.

Mr. Gomez had $15,000 to invest. Part of this money he invested at an annual rate of 6 % simple interest and the rest at an annual rate 7%. His interest income for the year was $980.

How much of his money was invested at 6%?
How much was invested at 7%?

The first step is to define the variables in the problem. To do this, we **always look at what we are being asked to find**. In this case, it is **"how much"** of the total **[$15,000]** is invested at each interest rate. So we define the variables as:

$$x = \text{the \textbf{amount} invested at 6\% \quad [0.06 as a decimal]}$$
$$y = \text{the \textbf{amount} invested at 7\% \quad [0.07 as a decimal]}$$

Now we need to develop two equations using these variables:

(1) an equation dealing with a total of $15,000. **[money invested]**
(2) an equation dealing with $980. **[the total interest earned]**

(1) The **first equation** is fairly straightforward:

x [the amount invested at 6%] + y [the amount invested at 7%] = \$15,000:

$$(1) \quad x + y = 15,000$$

(2) The second equation has to do with **interest earned**: We calculate this by multiplying the principal [the amount invested] by the interest rate ["annual" indicates a 1 year time period] Therefore, the second equation would be:

$$.06 \cdot x + .07 \cdot y = 980$$
$$\text{or just: } .06x + .07y = 980$$

We now have a system of two linear equations using the same variables. Therefore, there is **only one** value for x and y that will make **both equations true.** When we find those values, **we will have solved this problem.**

$$(1) \quad x + y = 15,000$$
$$(2) \quad .06x + .07y = 980$$

Decimal points confuse the issue, so we will multiply through equation (2) by 100: $100(.06x + .07y = 980) \Rightarrow$ (2) $6x + 7y = 98,000$

Inserting implied coefficients, the first equation looks like: (1) $1x + 1y = 15,000$

We see that arbitrarily multiplying equation (1) by -6 will be **convenient:** (1) $-6(1x + 1y = 15,000) \Rightarrow$ (1) $-6x + (-6y) = -90,000$

the second equation is: (2) $6x + 7y = 98,000$

Now, when we add the equations together, the **x** variable will "drop out": $1y = 8,000$

Or just: $y = 8,000$

[the amount invested at 7%]

So now we know that **\$8,000 was invested at 7%.** We use this information **in equation (1)** to find the value for x.

Let $y = 8,000$ in (1): $x + y = 15,000$

Substituting for **y**: $x + 8,000 = 15,000$

Adding $-8,000$ to both sides of the equation: $-8,000 \quad -8,000$

Results in: $x = 7,000$

[The amount invested at 6 %]

We have solved the problem:

We have $7,000 invested at 6% and $8,000 invested at 7%

Let's **verify** that this is indeed the solution.

| In the first equation: | (1) | $x + y = 15,000$ |
| If $x = 7,000$ and $y = 8,000$, then: | | $7,000 + 8,000 = 15,000$ |

These values for *x* and *y* have made the first equation true!

In the second equation:	(2) $.06x + .07y = 980$
If $x = 7,000$ and $y = 8,000$, then:	$.06 \cdot 7,000 + .07 \cdot 8,000 = 980$
Doing the multiplication:	$420 + 560 = 980$

These values for *x* and *y* have made the second equation true!

So we are confident that we have found the **unique** [only one] solution to this system of equations.

Example 3.

Suppose we have two alcohol solutions. The first is 30% alcohol and the second is 50% alcohol. **How much of each solution** is needed to produce **10 gallons** of a **solution** that is **42% alcohol**?

At first glance this problem appears to be quite complex. However, if we analyze it carefully and develop a **system of equations,** we can solve the problem.

First, we will **define the variables.** The key to defining the variables is to **look at what is being asked of us** ["How much of each solution?"].

So we   let x = **the number of gallons of 30% solution.**
and   let y = **the number of gallons of 50% solution.**

According to the information given, **the total must be 10 gallons of 42% solution.**

We know that we need 10 gallons of solution all together, so one linear equation will be:

(1) $x + y = 10$

The second equation is a little more difficult to conceptualize. We have to account for the fact that each solution that we are given has a different strength [30% & 50%], as well as our goal of making 10 gallons at a strength of 42% .

We see that 10 gallons of a 42% solution will contain **4.2 gallons** of **pure alcohol** [.42 x 10].

How do we end up with 4.2 gallons of pure (100%) alcohol given these two mixtures of 30% [.3 as a decimal] and 50% [.5 as a decimal]? The second equation will be based on **how much pure alcohol we need.**

We can represent the alcohol content of the x mixture as $.3 \cdot x$ or just $.3x$ and the alcohol content of the y mixture as $.5 \cdot x$ or just $.5x$. The total **pure alcohol content** must be 4.2 gallons [see above].

Therefore, our second equation, **based upon pure alcohol content** is:

$$(2) \quad .3x + .5y = 4.2$$

We now have a **system of two linear equations using the same variables.** Therefore, there is **only one value for x and y** that will solve **both** equations. When we find those values, **we will have solved this problem!**

$$(1) \quad x + y = 10 \quad \text{[gallons of mixture]}$$
$$(2) \quad .3x + .5y = 4.2 \quad \text{[gallons of pure alcohol]}$$

Inserting implied coefficients, the first equation looks like:

$$(1) \quad 1x + 1y = 10$$

To get rid of the decimals, we multiply through equation (2) by 10:

$$(2) \quad 10\left(.3x + .5y = 4.2\right) \quad \Rightarrow \quad (2) \quad 3x + 5y = 42$$

We see that arbitrarily multiplying equation (1) by -3 will be **convenient:**

$$(1) \quad -3\left(1x + 1y = 10\right) \quad \Rightarrow \quad (1) \quad -3x + \left(-3y\right) = -30$$

Adding the equations (1) and (2)

$$(2) \quad \underline{3x + 5y = 42}$$

causes the x variable to drop out [become "0"]:

$$2y = 12$$

Dividing both sides of the equation by 2:

$$\frac{2y}{2} = \frac{12}{2}$$

Results in:

$$y = 6$$

[the number of gallons of 50% mixture]

So now we know that there has to be **6 gallons of 50% mixture** [the y variable]. We use this information in equation (1) to find the x variable.

Let $y = 6$ in equation (1) $x + y = 10$.

Substituting for y: $x + 6 = 10$

Adding -6 to both sides of the equation: $\quad \underline{-6 \quad -6}$

Results in: $\quad x \quad = \quad 4$

[the number of gallons of 30% mixture]

So the **solution** is:

We need **4 gallons of 30% mixture** and **6 gallons of 50% mixture** to make **10 gallons of 42% mixture.**

To **verify** our solution, we know that we need 10 gallons altogether, **which is what we have** [4 gallons plus 6 gallons]. So far so good!

Now how about the amount of pure alcohol? We know that we need 4.2 gallons. Do we have it?

Let's look at our second equation: $\quad .3x + .5y = 4.2$

Substituting 4 for x and 6 for y: $\quad .3(4) + .5(6) = 4.2$

Results in: $\quad 1.2 + 3.0 = 4.2$

which satisfies the equation!

Voila! (as the French would say)

Concept Homework

As an assessment of your understanding of the concepts set forth in this section, answer the following questions. If necessary, review the material to help you arrive at the correct conclusions. If a statement is false, correct the wording to make it true.

1. True or False:
   a. In order to solve problems using a system of linear equations we need to develop 2 different equations that use the same variables in different ways.

   b. Graphing the solution lines of a system of linear equations will result in two intersection points provided that the lines are not parallel.

   c. We can arbitrarily multiply *either* equation [or both] in a system by any number as long as we multiply every term in the equation by that number. If we chose, we can multiply each equation by different numbers.

2. To solve "mixture problems", with alcohol solutions, we develop two equations with the same variables. One deals with the total amount of liquid. What must the other equation deal with?

3. If we are trying to determine how many of two different kinds of coins there are in someone's pocket, we need to know the total number coins. What other information would we need?

4. Make up your own "coin" problem starting with a solution and working backwards.

Solutions: 1a. True ; 1b. False; 1c. True; 2. The amount of pure alcohol. 3. The total value of the money.

Exercises

1. A man has only nickels and quarters in his pocket. There are a total of 23 coins. The total value of the coins is $3.15. How many nickels and how many quarters does he have?

2. The total income from two investments is $380.00. One investments yields 4% and the second yields 5%. How much was invested at each rate if the total investment was $8,000?

3. One solution contains 60% alcohol and the second contains 30% alcohol. How much of each solution is needed to make 33 liters that is 50% alcohol?

Solutions: 1. 13 nickels and 10 quarters; 2. $2,000 at 4% and $6,000 at 5%; 3. 22 liters at 60% and 11 liters at 30%.

Topic 16 - Laws of Exponents – Part I

When there is repeated multiplication of the same factor, it is convenient to express these operations with **exponential expressions**. In this topic, we will develop a definition of an exponential expression and learn how to use exponential expressions under the multiplication and division operations. We will also see how to raise an exponential expression to a power.

All the Laws of Exponents stem from the following basic definitions or properties:

A. <u>Exponential Expressions</u>: In general, x^a is an exponential expression. $"x"$ is called the **base** and $"a"$ is called the **exponent**. $"a"$ determines how many times x is used as a factor [multiplier].

If there is no exponent expressed, we assume that the exponent is "1". For example, y is equivalent to y^1 as an exponential expression. This follows from the definition above, since y^1 is the base y used as a factor once. Therefore $y^1 = y$.

Examples: $3^1 = 3$; $3^4 = 3 \cdot 3 \cdot 3 \cdot 3$ [which totals 81]; $x^3 = x \cdot x \cdot x$, and $3xy$ is equivalent to $3^1 x^1 y^1$.

B. *Any* **number raised to the power of zero has a value of 1.**
Ex 1. $3^0 = 1$ **Ex 2.** $523^0 = 1$ **Ex 3.** $(-2345)^0 = 1$ **Ex 4.** $x^0 = 1$
Ex 5. $3x^0 y^0 \Rightarrow 3 \cdot 1 \cdot 1 \Rightarrow 3$

C. The base of an exponent is considered to be the number or variable *immediately preceding* the exponent. If the base is meant to include more than that, a parenthesis will be put around what is to be included in the base.

Example 1. In the expression $3x^2$, the base is only x [the variable immediately preceding the exponent, 2.]. The coefficient, 3, is **not included** in the base. Therefore, $3x^2 = 3 \cdot x \cdot x$ However, in the expression $(3x)^2$, the coefficient 3 *is* **included** in the base and is being raised to the 2nd power along with the $"x"$. Therefore, $(3x)^2 \Rightarrow 3x \cdot 3x \Rightarrow 3 \cdot 3 \cdot x \cdot x \Rightarrow 9x^2$.
[Note: multiplication is commutative. Therefore, the factors can be arranged in any order.]

Example 2. In the expression -3^2, the negative symbol is **not** included in the base. Therefore: $-3^2 \Rightarrow -(3 \cdot 3) = -9$. However, in the expression $(-3)^2$, the negative symbol **is** included in the base. Therefore: $(-3)^2 \Rightarrow (-3)(-3) = 9$.

__Multiplying Exponentials__ [The Product Rule]: $a^m \cdot a^n = a^{m+n}$

When we multiply bases that are _exactly_ the same, we add the exponents.
   Ex 1. $2^2 \cdot 2^3 \Rightarrow 2^{(2+3)} \Rightarrow 2^5 \Rightarrow 2 \cdot 2 \cdot 2 \cdot 2 \cdot 2 = 32$

We can see the logic behind the rule by expanding each exponential before doing the multiplication: $2^2 = 2 \cdot 2$ and $2^3 = 2 \cdot 2 \cdot 2$. So multiplying them together results in $2 \cdot 2 \cdot 2 \cdot 2 \cdot 2$ or 2^5. Therefore, we can arrive at the same result by **adding** the exponents.

   Ex 2. $x^3 \cdot x^6 \Rightarrow x^{3+6} = x^9$

   Ex 3. $p^3 \cdot q^4$ Cannot be simplified . **[the bases are not identical!]**

   Ex 4. $x^3 + x^5$ Cannot be simplified. **[the _operation_ is not multiplication!]**

__Dividing Exponentials__: $\dfrac{a^m}{a^n} = a^{m-n}$

When we divide exponentials, we subtract the denominator's exponent from the numerator's exponent. Note: The bases must _exactly_ be the same to perform this operation.

   Ex 1. $\dfrac{2^3}{2^2} \Rightarrow 2^{3-2} = 2^1$ or 2

Again, we can see the logic behind the rule by expanding each exponential before doing

the division: $\dfrac{2^3}{2^2} \Rightarrow \dfrac{2 \cdot 2 \cdot 2}{2 \cdot 2} \Rightarrow \dfrac{\cancel{2} \cdot \cancel{2} \cdot 2}{\cancel{2} \cdot \cancel{2}} \Rightarrow \dfrac{2}{1}$ or just 2 . Therefore, we can arrive at the

same result by subtracting the denominator's exponent from the numerator's exponent..

   Ex 2. $\dfrac{x^3}{x} \Rightarrow \dfrac{x^3}{x^1} \Rightarrow x^{3-1} = x^2$

   Ex. 3. $\dfrac{a^6 b^3 c^4}{a^6 b^2 c^2} \Rightarrow a^{6-6} b^{3-2} c^{4-2} \Rightarrow a^0 b^1 c^2 \Rightarrow 1bc^2$ or bc^2

   Ex. 4. Suppose the exponent in the denominator (n) is larger than

   the exponent in the numerator (m). For example, $\dfrac{a^3}{a^4} = a^{3-4} = a^{-1}$

It is possible to have negative exponents. We will discuss them in the Topic 18.

Raising a power to a power [The power rule]: $\boxed{\left(a^m\right)^n = a^{m\cdot n}}$

When raising an exponential expression to a power, we multiply the exponents.

Ex 1. $\left(2^2\right)^3 \Rightarrow 2^{2\cdot3} \Rightarrow 2^6 \Rightarrow 2\cdot2\cdot2\cdot2\cdot2\cdot2 = 64$

Once again, we can see the logic behind the rule by considering 2^2 as a base raised to the exponent of 3. By definition this can be viewed as $2^2 \cdot 2^2 \cdot 2^2$ which we know from the product rule to be 2^6. Therefore, we can arrive at the same result by multiplying the exponents.

Ex 2. $\left(x^2\right)^4 = x^{2\cdot4} = x^8$

When the base of an exponential is made up of factors: $\boxed{\left(ab\right)^m = a^m b^m}$

The parenthesis in this rule indicates that both a and b are part of the base of the exponential expression. Therefore, each of these factors are raised to the power indicated.

Ex 1. $\left(xy\right)^2 = x^2 y^2$ [Note: *Both* factors in the base are raised!]

Ex 2. $\left(-3a\right)^2 \Rightarrow \left(-3\right)^2 \cdot \left(a^1\right)^2 = 9a^2$

Ex 3. $-\left(3a\right)^2 \Rightarrow -\left(3^1 \cdot a^1\right)^2 \Rightarrow \left(-1\right)\cdot3^2 a^2 = -9a^2$

Note: When the negative symbol is outside of the parentheses, it is not treated as part of the base. In effect, the exponential is being multiplied by -1. Therefore the result of this expression will be negated.

Ex 4. $\left(2x^4\right)^5 \Rightarrow \left(2^1 \cdot x^4\right)^5 \Rightarrow 2^5 x^{20} = 32x^{20}$

Ex 5. $\left(4a^3b^3\right)^2 \Rightarrow 4^2 \cdot a^{3\cdot2} \cdot b^{3\cdot2} = 16a^6b^6$

Raising a fractional base to a power: $\left(\dfrac{x}{y}\right)^a = \dfrac{x^a}{y^a}$

When raising a fractional base to an exponent, both the numerator and the denominator are raised to that power.

Ex 1. $\left(\dfrac{2}{3}\right)^2 \Rightarrow \dfrac{2^2}{3^2} = \dfrac{4}{9}$

Ex 2. $\left(\dfrac{x}{5}\right)^3 = \dfrac{x^3}{5^3} = \dfrac{x^3}{125}$

Ex 3. $\left(\dfrac{-2p^2}{xy^3}\right)^3 \Rightarrow \dfrac{\left(-2\right)^3\left(p^2\right)^3}{x^3 y^9} = \dfrac{-8p^6}{x^3 y^9}$ or $-\dfrac{8p^6}{x^3 y^9}$

148

Concept Homework

If an answer is true, explain why. If it is false, explain why and show how you would change the statement to make it true.

True or False:

1. $\dfrac{1}{y^0} = 1$

2. If a variable is written without a power, it has no power.

3. When multiplying two exponential terms whose bases are not the same, we must multiply bases as well as the exponents. For example, $x^4 \cdot y^3 = xy^{12}$.

4. When dividing exponentials whose bases are the same, we can eliminate identity elements to simplify the fraction.

For Example: $\dfrac{y^5}{y^2} \Rightarrow \dfrac{\cancel{y^2} y^3}{\cancel{y^2}} = y^3$

5. $6^3 \Rightarrow (3 \cdot 2)^3 = 3^3 \cdot 2^3$

6. x^0 is the identity element for multiplication.

Solutions: 1. True; 2. False; 3. False; 4. True; 5. True; 6. True;

Exercises

1. Perform the indicated operations and simplify:

  a. $2^2 \cdot 2^3$ **b.** $-2^2 \cdot 2^3$ **c.** $a \cdot a^4$ **d.** $(-c)^4$ **e.** $x^2 + x^6$ **f.** $a^2(-b)^4$

2. Perform the indicated operations and simplify:

  a. $\left(-2^2\right)^3$ **b.** -3^2 **c.** $(-2)^4$

3. Perform the indicated operations and simplify:

  a. $\dfrac{x^5}{x^2}$ **b.** $\dfrac{6a^4b^8}{ab^7}$ **c.** $\dfrac{(-4)^5}{(-4)^3}$ **d.** $\left(\dfrac{2x^3}{3y^5}\right)^2$ **e.** $\dfrac{(2a^2b^3)^2}{-4a^4b}$

Solutions: **1a.** 32 ; **1b.** −32 ; **1c.** a^5 ; **1d.** c^4 ; **1e.** cannot be simplified; **1f.** a^2b^4 ; **2a.** −64 ; **2b.** −9 ; **2c.** 16 ; **3a.** x^3 ; **3b.** $6a^3b$;

  3c. 16 ; **3d.** $\dfrac{4x^6}{9y^{10}}$; **3e.** $-b^5$.

Topic 17 - Operations with Polynomials: Addition, Subtraction & Multiplication

Prior to this topic, we have been using polynomials without having defined them mathematically. As polynomials are the basic form of algebraic expressions, we should know how they are defined. We have had long experience with operating with arithmetic numbers. We must also know how to **operate** with polynomials. Operations include **addition, subtraction, multiplication, division, and exponential operations**. In this topic we will deal with **addition subtraction and multiplication** of polynomials.

Definitions

A *term* can simply be a number. It can also be considered a **monomial** [See the definition below]. Terms can also have numbers that are multiplying variable(s)

Examples: 4, $2x$, $3r^2$, $-7pq^2r^5$.

Monomials are a **sub-category of polynomials** involving **exactly one term**.

Notice that $2x \Rightarrow 2 \cdot x$; $\frac{1}{3}r^2 \Rightarrow \frac{1}{3} \cdot r^2$; $-7pq^2r^5 \Rightarrow -7 \cdot p \cdot q^2 \cdot r^5$. In a monomial there can be both multiplication and division [a fraction implies a division operation] as well as exponents as long as the exponents of the variables are positive.

A *polynomial* [that is not a monomial] is a series of **terms** connected to each other by **addition** or **subtraction**.

Binomials are a **sub-category of polynomials** with **exactly two terms** connected by **addition** or **subtraction**.

Trinomials are a **sub-category of polynomials** with **exactly three terms** connected by **addition** or **subtraction**.

Polynomials can only have **variables with positive exponents**. For example, $2x^{-2} + 3x$ is **not** considered a polynomial since there is a negative exponent involved. $2x^{-2} \Rightarrow \frac{2}{x^2}$. There cannot be a variable in the denominator of a polynomial as it would no longer conform to the definition of a polynomial. [We will discuss negative exponents in Topic 18]

A *constant* [stand-alone number without a variable] also falls into the category or a **term** or **monomial.**

Generally speaking **monomials and polynomials have variables**. However, a constant, for example 5, is equivalent to $5x^0$ or $5 \cdot 1$ which we know from our study of the laws of exponents. So, we see that a constant can be expressed as an **equivalent** monomial with a variable involved. a variable raised to the zero power will not change the value of its coefficient!

"Poly" in Latin means "many", so it's not surprising that a **polynomial** is an algebraic expression with **one or more monomials** *connected by addition or subtraction.*

Examples:

$$2x^2, \quad 3y^2 + 2xy, \quad 4x^3 - 3x - 2, \quad 2x^2 + 3xy + 2y^2 - 5$$

[monomial]   [binomial]   [trinomial]   [four-termed polynomial]

Although we have given some polynomials special classifications as shown above **[monomial (one term), binomial (two terms), trinomial (three terms)]**, they all belong to the more general **polynomial class**. But as we previously mentioned, Polynomials can never have **negative exponents.**

The *degree of a term* is the sum of the exponents of the variables in that term. For example, $3y^3$ has a degree of 3. However, $3xy^3 \Rightarrow 3x^1y^3$ has a degree of 4 since we must add the exponents of the variables within the term to determine its degree. A constant, 5 for example, has a degree of zero since 5 can be viewed as its equivalent, $5x^0$.

The *degree of a polynomial* is the greatest degree of any of the terms in the polynomial. For example, the polynomial $3x^2y + 2x - 1$ would have a degree of 3. The greatest degree of the 3 terms is that of the first term, $3x^2y \Rightarrow 3x^2y^1$ which has a degree of 3. This is the greatest degree of all of the polynomial's terms.

Addition of Polynomials [also referred to as "combining like terms"]

Monomials with **like terms** have **exactly the same variables raised to exactly the same powers**. *Only the coefficients* may differ.

Examples of like terms: (a) $3x^2y$ and $-2x^2y$;   (b) $12x^3yz^4$ and $3x^3z^4y$

Notice that in both of these examples, the variable parts of the terms are **exactly** the same, including their exponents. In example (b), the variables are arranged in a **different order**. However, since monomials represent a series of values being multiplied, they are *still* identical since order doesn't matter with the multiplication operation **[Multiplication is commutative!]**. Also, notice that **the coefficients do not have to be alike.**

Examples of *unlike* terms: (a) $12x^3yz^4$ and $-3x^3z^4$ (b) $3x^2y$ and $3xy^2$

Notice that in the example (a), the first term has a y variable and the second term does not. In example (b), the variables are **not** all raised to the same power in each term.]

Only like terms may be added. This is done by adding *only the coefficients. The variable part of the like terms remain the same in the resulting polynomial.* An analogy to real life is trying to add apples and oranges. They are *unlike* and therefore cannot be combined through addition. For example, 3 apples and 4 oranges cannot be combined through addition. They remain 3 apples and 4 oranges. However, 3 apples and 4 apples are **"like"** and can be combined through addition to become 7 apples.

Examples: (a) $12x^3yz^4 + 3x^3yz^4 = 15x^3yz^4$

[We usually order the variables alphabetically.]

(b) $3x^2y - 2yx^2 = 1x^2y$ [or just x^2y] [These are like terms as the order of the variables doesn't matter.]

(c) $4x^2y + 5xy^2 \Rightarrow 4x^2y + 5xy^2$ [cannot be combined]

Adding several like terms in a polynomial

Example: **Add the following polynomials:** $\left(3x^3 + 3x^2 + 9x + 8\right) + \left(-7x^2 - 8x + 1\right)$

The polynomials are parenthesized to distinguish one from the other. However, because we are adding them together, we can rearrange the terms into "like" groups:

[Addition is commutative. We can reorder them as we wish.]

$$3x^3 + 3x^2 \quad +9x + \ 8$$
$$\underline{\quad\quad -7x^2 \quad -8x \quad\quad 1}$$

Adding like terms: $\quad 3x^3 - 4x^2 + 1x + 9$

This is conventionally written: $\quad 3x^3 - 4x^2 + x + 9$

[not showing the "1" coefficient of the x term.]

Subtracting Polynomials

Example 1. **Subtract the following polynomials:** $\quad \left(4x^4 - 3\right) - \left(2x^4 - 8x^2 + 2\right)$

We want to change the subtraction sign **between** the parentheses to addition. We do this by reversing the sign of **every term**

the parenthesis following the subtraction symbol: $\quad \left(4x^4 - 3\right) + \left(-2x^4 + 8x^2 - 2\right)$

As we are now **adding** the polynomials, we group "like terms"

together:
$$4x^4 \quad\quad\quad -3$$
$$\underline{-2x^4 + 8x^2 - 2}$$

resulting in:
$$2x^4 + 8x^2 - 5$$
[Solution]

[We list the highest degree term first and then in descending order.]

Don't let the parentheses in these examples fool you into thinking that there is multiplication involved. The addition or subtraction sign **between the parentheses** excludes multiplication. The parentheses are merely grouping terms of different polynomials together.

Multiplying Polynomials

As you would expect, **the multiplication operation with polynomials has different rules than the addition operation, just as in arithmetic.** *Both* the coefficients *and* the variables are multiplied.

Example 1. Multiply: $-2x \cdot 8x \implies -2 \cdot x^1 \cdot 8 \cdot x^1 \implies -2 \cdot 8 \cdot x^1 x^1 = -16x^2$
[solution]

Notice that we gave the variables the implied power of "1" and re-arranged the factors to group coefficients together and variables together, utilizing the commutative property of multiplication! We are also employing the **Product Rule** of exponents:
"**When multiplying exponentials with the same base, we keep the base and add the exponents.**"

Example 2.   **Multiply:** $\left(3p^2q\right)\left(5p^3 - 2pq + q^3\right)$

The multiplication here needs to be **distributed**. First, we
  should give every term a coefficient and every variable
    its implied power [**if they do not already have one**]: $\left(3p^2q^1\right)\left(5p^3 - 2p^1q^1 + 1q^3\right)$

We are now ready to **distribute**. We must multiply
the term in the first parenthesis by **every** term in
    the second parenthesis: $\left(3p^2q^1\right)\left(5p^3\right) + \left(3p^2q^1\right)\left(-2p^1q^1\right) + \left(3p^2q^1\right)\left(1q^3\right)$
[It is a good idea to actually write this step out rather than do it in your head to eliminate inadvertent errors.]

We multiply coefficients *and* variables: $\left(15p^5q^1\right) + \left(-6p^3q^2\right) + \left(3p^2q^4\right)$

 This is more conventionally expressed as: $15p^5q^1 - 6p^3q^2 + 3p^2q^4$
[solution]

Since there are no "like-terms" to be combined through addition, we are finished.

Multiplying Binomials using FOIL (First, Outside, Inside, Last)

Example 1.   **Multiply:** $\left(8x - 3\right)\left(2x - 1\right)$
  [Notice that both polynomials are binomials since they each have two terms.]

                    Now comes the '**FOIL**':
        First: Multiply the first term in each parenthesis: $8x^1 \cdot 2x^1 = \quad 16x^2$
        Outside: Multiply first term [**first parenthesis**] by
                the last term [**second parenthesis**]: $8x(-1) = \quad -8x$
        Inside: Multiply last term [**first parenthesis**] by
                the first term [**second parenthesis**]: $-3(2x) = \quad -6x$
        Last: Multiply last term [**first parenthesis**] by
                the last term [**second parenthesis**]: $(-3)(-1) = \quad 3$

The resulting polynomial is: $\qquad$ $16x^2 - 8x - 6x + 3$

Combining "like" middle terms: $\qquad$ $16x^2 - 14x + 3$

[solution]

An alternative to the FOIL method is to form a **matrix** with each term represented as shown below with one polynomial listed horizontally and one vertically. We multiply every term in the rows by every term in the columns:

Multiply	$2x^1$	-1
$8x^1$	$16x^2$	$-8x^1$
-3	$-6x^1$	3

We now add all the terms in the shaded boxes, combining the like terms [$-6x^1$ and $-8x^1$], resulting in: $\qquad$ $16x^2 - 14x^1 + 3$

Or just: $\qquad$ $16x^2 - 14x + 3$

[solution]

Multiplying Conjugates and Squaring Binomials

Example 1. $\qquad\qquad\qquad$ **Multiply:** $\quad (2x + 3y)(2x - 3y)$

Notice that the terms in the two binomials are the same **except** that the first binomial has the addition operation and the second has the subtraction operation. These arrangements are called **conjugates**. They have special significance for **factoring** which we will discuss in a later topic. Writing in the implied "1" power for the variables:

$(2x^1 + 3y^1)(2x^1 - 3y^1)$

When we **FOIL** the binomials: $\quad$ **First:** $\quad 2x^1 \cdot 2x^1 = 4x^2$

Outside: $\quad (2x)(-3y) = -6xy$

Inside: $\quad 3y \cdot 2x = 6xy$

[We listed the variables alphabetically.]

Last: $(3y^1)(-3y^1) = -9y^2$

Adding the terms together, we get: $\quad 4x^2 - 6xy + 6xy - 9y^2$

When combining like terms, we see that **the middle terms are opposites!**

$[-6xy + 6xy \Rightarrow 0xy \Rightarrow 0]$

Opposites add to zero, so the result is: $\qquad 4x^2 - 9y^2$

When we multiply *any* conjugates, the middle term always drops out!

Using the **matrix method**:

Multiply	$2x^1$	$3y^1$
$2x^1$	$4x^2$	$6x^1y^1$
$-3y^1$	$-6x^1y^1$	$-9y^2$

We now add all the terms in the shaded boxes: $4x^2 + 6x^1y^1 - 6x^1y^1 - 9y^2$

Combining the like terms: $4x^2 + 0x^1y^1 - 9y^2$

The middle terms are opposites!

Opposites add to zero, so the result is: $4x^2 - 9y^2$

[solution]

We see that multiplying conjugates results in a *binomial*. There is no "middle term".

Example 2.   **Expand the Exponential** [do the multiplication]: $(x+y)^2$

We would be tempted to think that the result would be $x^2 + y^2$.

However, this is **incorrect as we will see below.**

As $(x+y)^2$ is an **exponential** expression with a base of $(x+y)$,

Raising the base to the 2nd power requires us to use

The base as a factor twice. Therefore $(x+y)^2 \Rightarrow$ $(x+y)(x+y)$

To multiply we can **FOIL** these binomials.

We can start by adding the implied coefficients and powers: $(1x^1 + 1y^1)(1x^1 + 1y^1)$

FOILing, results in:

$1x^2 + 1x^1y^1 + 1x^1y^1 + 1y^2$

Combining like terms: $1x^2 + 2x^1y^1 + 1y^2$

Or just: $x^2 + 2xy + y^2$

[Solution]

Notice that $(x+y)^2$ **is NOT equivalent to** $x^2 + y^2$. It expands to $x^2 + 2xy + y^2$

In fact, the only time that we do not have a middle term when multiplying binomials is when they are conjugates!

[Conjugates are identical binomials except that one has an addition operation and the other a subtraction operation as in Example 1.]

Multiplying a Binomial and a Trinomial

The FOIL method only works when **multiplying two binomials**. Therefore, with larger polynomials, we could use the **matrix method**:

Example 1.   **Multiply:** $\left(2x^2 - 5\right)\left(3x^3 - x^2 - 6x\right)$

To avoid confusion we add the "missing" coefficient

for the x^2 term and the implied power of the $6x$ term: $\left(2x^2 - 5\right)\left(3x^3 - 1x^2 - 6x^1\right)$

Once we have done this, we need to multiply every term in the binomial by every term in the trinomial. This can be accomplished by a **multiplication matrix** with the binomial represented in the rows and the trinomial in the columns:

Multiply	$3x^3$	$-1x^2$	$-6x^1$
$2x^2$			
-5			

Now we just fill in the boxes:

Multiply	$3x^3$	$-1x^2$	$-6x^1$
$2x^2$	$6x^5$	$-2x^4$	$-12x^3$
-5	$-15x^3$	$5x^2$	$30x^1$

Now we add all the terms **in the shaded boxes**,

while listing them by the size of their degree: $6x^5 - 2x^4 \underline{-12x^3 - 15x^3} + 5x^2 + 30x^1$

Combining like terms **[the only like terms are the x^3's]** : $6x^5 - 2x^4 - 27x^3 + 5x^2 + 30x$

[Notice that we dropped the power of "1" on the $30x$ term as we no longer need to express it.]   [solution]

Another method of multiplying polynomials is to set up the terms as if we were multiplying multi-digit regular numbers:

Using the **same polynomials as in the previous example**:

$$3x^3 - 1x^2 - 6x^1$$
$$\text{x} \quad \underline{\quad 2x^2 \quad -5 \quad}$$

We start by multiplying every term in the first row $\left[3x^3 - 1x^2 - 6x^1\right]$, by the **first term** in the second row $\left[2x^2\right]$: $\quad 6x^5 - 2x^4 - 12x^3$

Now, we multiply the same first row $\left[3x^3 - 1x^2 - 6x^1\right]$,

by the **second term** in the second row $[-5]$: $\quad \underline{-15x^3 + 5x^2 + 30x}$

Notice that we group like terms in the same columns

to make it easier to add them together: $\quad 6x^5 - 2x^4 - 27x^3 + 5x^2 + 30x$

[solution]

Concept Homework

As an assessment of your understanding of the concepts set forth in this section, answer the following questions. If necessary, review the material to help you arrive at the correct conclusions. With "true or false" statements, if the statement is true, make up an example that supports the statement without using examples given in the material. If a statement is false, correct the wording to make it true. Then make up an example that supports the statement without using examples given in the material.

<u>**True or False**</u>:

1. A binomial is an algebraic expression that has two terms connected by the addition or subtraction operation.

2. Like terms have to have identical variables, including their powers. However, the variables can be arranged in a different order and still be "like". The commutative property of multiplication permits this.

3. When *adding* terms in a polynomial, they must be "like". The only part that gets added are the coefficients. The variables parts remain the same after they are added.

4. When *multiplying* polynomials, we multiply the coefficients and keep the variables as they were originally.

5. We can only *multiply* polynomials if they consist of "like terms".

6. Binomials are polynomials with two terms. "FOILing" refers to the way we multiply the terms of each binomial.

7. Conjugates are binomials that are identical.

8. When we multiply conjugates, the result will *not* have a "middle term".

9. When we square a binomial [raising it to the 2nd power] there will always be a "middle term".

10. $\left(m-n\right)^2 = m^2 - n^2$

11. $\left(m+n\right)\left(m-n\right) = m^2 - n^2$

12. $\left(m+n\right)^2 = m^2 + 2mn + n^2$

Solutions: 1. True; 2. True; 3. True; 4. False; 5. False; 6. True; 7. False; 8. True; 9. True; 10. False; 11. True; 12. True

158

Exercises

1. Add the following polynomials:

a. $\left(4x^2 - 2x\right) + \left(2x^3 - 3x^2 + 4x - 3\right)$ **b.** $\left(x^2yz + 2xyz^2 - 3xy^2z\right) + \left(-5xy^2z + x^2yz - 2xyz^2\right)$

2. Subtract the following polynomials:

a. $\left(4zy + 3y - 2xz\right) - \left(2yz + 2z - 3xz + 5xyz\right)$ **b.** $\left(x^3 - 2x^2 + 3x - 4\right) - \left(3x^2 + 4x - 4\right)$

3. Multiply the following polynomials:

a. $x^4\left(x^3 - x + 2\right)$ **b.** $5a^3b^2\left(ab^2 - bc + 4a\right)$

4. Multiply the following binomials using both the FOIL method and matrix method:

a. $\left(9x - 2\right)\left(4x - 3\right)$ **b.** $\left(2 - 9x\right)\left(3 + x\right)$ **c.** $\left(2x^2 - 4x\right)\left(5x^3 + 3\right)$ **d.** $\left(2x - 5\right)\left(2x + 5\right)$
e. $\left(2x - 5\right)^2$

5. Multiply the following polynomials using two different methods:

a. $\left(2x^2 - 3x + 6\right)\left(x^2 + 2x + 4\right)$ **b.** $\left(3x^2 - x + 2\right)\left(5x - 3\right)$

6. Perform the following operations making sure that follow "order of operations" rules:

$$\left(2x + 3\right)^2 - \left(2x - 3\right)^2$$

Solutions: **1a.** $2x^3 + x^2 + 2x - 3$; **1b.** $2x^2yz - 8xy^2z$; **2a.** $-5xyz + 2yz + xz + 3y - 2z$ [**The order of these terms is in descenending order by degree.**] ; **2b.** $x^3 - 5x^2 - x$; **3a.** $x^7 - x^5 + 2x^4$; **3b.** $5a^4b^4 - 5a^3b^3c + 20a^4b^2$; **4a.** $36x^2 - 35x + 6$; **4b.** $-9x^2 - 25x + 6$ [**The order of these terms is in descenending order by degree.**] ; **4c.** $10x^5 - 20x^4 + 6x^2 - 12x$; **4d.** $4x^2 - 25$; **4e.** $4x^2 - 20x + 25$;
5a. $2x^4 + x^3 + 8x^2 + 24$; **5b.** $15x^3 - 14x^2 + 13x - 6$; **6.** $24x$

Topic 18 – Laws of Exponents – Part II – Negative Exponents and Polynomial Division

We have previously learned the **quotient rule** for dividing same base exponentials:

$$\frac{a^m}{a^n} = a^{m-n}.$$

We haven't discussed what happens when a^n in the denominator has a larger exponent than a^m in the numerator.

For **example**, how would we evaluate: $\dfrac{x^2}{x^4}$

Following the above quotient rule, this would result in: $x^{2-4} \Rightarrow x^{-2}$

As can be seen, such an evaluation would yield a **negative exponent!**

We can view this example in another way: $\dfrac{x^2}{x^4} = \dfrac{\overset{1 \cdot 1}{\cancel{x} \cdot \cancel{x}}}{\underset{1 \cdot 1}{\cancel{x} \cdot \cancel{x} \cdot x^1 \cdot x^1}} = \dfrac{1}{x^2}$

[Two of the *x's* in the numerator and denominator "cancel" each other!]

We have the same original example equaling both x^{-2} and $\dfrac{1}{x^2}$.

Therefore, **both forms of the simplification must be equal:** $x^{-2} = \dfrac{1}{x^2}$

We can generalize this idea as a **definition of a negative exponent:** $\dfrac{x^{-a}}{1} = \dfrac{1}{x^a}$

or its corollary : $\dfrac{1}{x^{-a}} = \dfrac{x^a}{1}$

We justify this corollary, $\dfrac{1}{x^{-a}} = \dfrac{x^a}{1}$, as follows:

Using our definition of a negative exponent: $\dfrac{1}{x^{-a}} \Rightarrow \dfrac{1}{\frac{1}{x^a}}$

Simplifying the complex fractions: $\dfrac{1}{\frac{1}{x^a}} \Rightarrow 1 \div \dfrac{1}{x^a} \Rightarrow 1 \cdot \dfrac{x^a}{1}$ or just $\dfrac{x^a}{1}$

One very useful result of this analysis is that a negative exponent can be changed to positive simply by moving it from the numerator of a fraction to the denominator or from the denominator to the numerator depending on where the negative exponent originated.

Simplifying Negative Exponents [changing their exponents from negative to positive]:

[Note: The sign of the exponent is reversed when the exponential is moved from numerator to denominator or denominator to numerator.]

Ex 1. $9^{-2} \Rightarrow \dfrac{9^{-2}}{1} \Rightarrow \dfrac{1}{9^2} = \dfrac{1}{81}$

Ex 2. $n^{-5} \Rightarrow \dfrac{n^{-5}}{1} = \dfrac{1}{n^5}$

Ex 3. $\dfrac{2}{x^{-3}} = \dfrac{2x^3}{1}$ or just $2x^3$

Ex 4. $\dfrac{x^{-2}}{y^{-3}} = \dfrac{y^3}{x^2}$ [changing negative to positive exponents by putting them on the opposite side of a fraction.]

Ex. 5. $\dfrac{x^{-4}y^3z^{-5}}{x^2y^{-2}z^6} \Rightarrow \dfrac{y^3y^2}{x^4x^2z^5z^6} \Rightarrow \dfrac{y^{3+2}}{x^{4+2}z^{5+6}} \Rightarrow \dfrac{y^5}{x^6z^{11}}$

Ex 6. $-(3y)^{-1} \Rightarrow (-1)\cdot\dfrac{(3y)^{-1}}{1} \Rightarrow (-1)\cdot\dfrac{1}{(3y)^1} = -\dfrac{1}{3y}$.

[Note: When the negative symbol is outside of the parentheses, it is not treated as part of the base. In effect, the exponential is being multiplied by -1. Therefore the result of this expression will be negated.]

Dividing Polynomials

To begin our discussion of dividing polynomials it will be helpful to review the arithmetic of adding fractions with like denominators. Recall that if the denominators of fractions are the same, we simply add the numerators.

Suppose we are adding: $\dfrac{9}{3}+\dfrac{6}{3}+\dfrac{12}{3}$.

Since the denominators are the same, we add the numerators and keep the denominator: $\dfrac{9+6+12}{3}$

As the two expressions above are equivalent, they are **interchangeable**. Therefore: $\dfrac{9+6+12}{3} = \dfrac{9}{3}+\dfrac{6}{3}+\dfrac{12}{3}$

This analysis lends itself to understanding how to **divide polynomials**.

Example 1.   **Divide:** $\left(9x^4 - 6x^3 + 12x^2\right) \div -3x^2$

Since every division problem can be represented

as a fraction we can transform this problem into:

$$\frac{9x^4 - 6x^3 + 12x^2}{-3x^2}$$

We can change subtraction to addition

[in this case, in the numerator]:

$$\frac{9x^4 + \left(-6x^3\right) + 12x^2}{-3x^2}$$

Just as in the previous arithmetic example, we can break the fraction down into 3 separate fractions

with the same denominator:

$$\frac{9x^4}{-3x^2} + \frac{-6x^3}{-3x^2} + \frac{12x^2}{-3x^2}$$

We can simplify each fraction individually employing

the "**quotient rule**" of exponents to the **first fraction**:

$$\frac{9x^4}{-3x^2} \Rightarrow -3x^{4-2} = -3x^2$$

Now to the **second fraction**:

$$\frac{-6x^3}{-3x^2} \Rightarrow 2x^{3-2} = 2x^1$$

Now to the **third fraction**:

$$\frac{12x^2}{-3x^2} \Rightarrow -4x^{2-2} = -4x^0$$

Adding these monomials results in: $-3x^2 + 2x^1 - 4 \cdot 1$ $\left[x^0 = 1\right]$

Or more conventionally: $-3x^2 + 2x - 4$

[Solution]

Example 2.

As we know from our study of arithmetic, sometimes division leaves a remainder if the divisor (denominator) doesn't divide the dividend (numerator) evenly. Here is such an example with polynomials:

Divide: $\dfrac{30x^7 + 10x^2 + 3x}{5x^2}$

First we break the fraction into 3 separate fractions: $\dfrac{30x^7}{5x^2} + \dfrac{10x^2}{5x^2} + \dfrac{3x}{5x^2}$

Employing the "**quotient rule**" of exponents to the

first fraction: $\dfrac{30x^7}{5x^2} \Rightarrow 6x^{7-2} = 6x^5$

Now to the **second fraction**: $\dfrac{10x^2}{5x^2} \Rightarrow 2x^{2-2} \Rightarrow 2x^0 = 2$

[In the above fraction, $\dfrac{x^2}{x^2}$ "cancels out".]

The denominator of the **third fraction** does not divide the numerator evenly, so we simplify it and leave it in

fractional form: $\dfrac{3x}{5x^2} \Rightarrow \dfrac{3x^{1-2}}{5} = \dfrac{3x^{-1}}{5}$

Using our **negative exponents rule**, we can change the negative exponential to positive by taking it from the numerator to the denominator: $\dfrac{3}{5x^1} \Rightarrow \dfrac{3}{5x}$

[You can look at this simplification in another way: $\dfrac{3x}{5x^2} \Rightarrow \dfrac{3x}{5x \cdot x} = \dfrac{3}{5x}$.

One x in the numerator "cancels" one in the denominator.]

Adding the results, we get: $6x^5 + 2 + \dfrac{3}{5x}$

[solution]

[The last term remains as a fraction since the denominator does not divide the numerator evenly.]

Concept Homework

As an assessment of your understanding of the concepts set forth in this section, answer the following questions. If necessary, review the material to help you arrive at the correct conclusions. With "true or false" statements, if the statement is true, make up an example that supports the statement without using examples given in the material. If a statement is false, correct the wording to make it true. Then make up an example that supports the statement without using examples given in the material.

True or False

1. An exponential with a negative exponent cannot always be expressed as an equivalent exponential with a positive exponent.

2. When dividing exponentials whose bases are the same, the division is eliminating identity elements to simplify the fraction.

3. $\left(\dfrac{3^{-2}}{4^{-2}}\right)^{3} = \dfrac{4^{6}}{3^{6}}$

4. The division of the polynomial $3x^{2} - 2x + 5$ by the divisor, $3x$, can be expressed as 3 separate fractions in order to perform the division and simplify each term in the polynomial. The result would be: $x - \dfrac{2}{3} + \dfrac{5}{3x}$.

Solutions: 1. False; 2. True; 3. True; 4. True

Exercises

Simplify. Write your solutions with positive exponents:

1. $\dfrac{2^{-1}}{2^2}$ **2.** $\dfrac{3^{-9}}{3^{-6}}$ **3.** $\dfrac{x^{-2}y^4}{x^2y^{-4}}$ **4.** $\dfrac{\left(xy^{-3}\right)^{-2}}{\left(x^{-2}y\right)^3}$

[Hint: Expand both the numerator and denominator first.]

5. $\dfrac{6x^2y^2+4x^2y-2xy}{2xy}$ **6.** $\dfrac{12a^3b^2-4a^2b+a}{2a^2b}$

7. $\dfrac{\left(a+b\right)^2-\left(a-b\right)^2}{2ab}$

[Hint: Expand the numerator and combine like terms. Then perform the division.]

Solutions: **1.** $\dfrac{1}{8}$; **2.** $\dfrac{1}{27}$; **3.** $\dfrac{y^8}{x^4}$; **4.** x^4y^3 ; **5.** $3xy+2x-1$; **6.** $6ab-2+\dfrac{1}{2ab}$; **7.** 2 .

Topic 19 - Factoring Polynomials – Part I

Solving equations is a primary goal in algebra. The **factoring process** is useful in reaching this goal as it allows us to write a polynomial in an equation as an equivalent product of simpler polynomials. In the previous topic we learned how to **multiply** polynomials through **distribution**. Factoring **reverses** the distributing process, so it is a **division** process. We will start the discussion by reviewing an example of the distributing process.

Example 1.   **Distribute**: $4x(2x+3)$

We multiply the **monomial,** $4x$, by *every* term in the **binomial**: $8x^2 + 12x$

Factoring is **reversing** the process!

Example 2.   **Factor**: $8x^2 + 12x$

A **factor**, or, more precisely, a ***common factor,*** is a term that will **divide evenly** into *all* the terms presented. A ***greatest common factor*** is the **largest** such factor. We look for common factors: **those terms that will divide both** $8x^2$ **and** $12x$ **evenly.**

First, look at the **coefficients** of each term: 8 and 12. Although 2 would divide both coefficients evenly, and therefore would be a **common factor,** it would **not** be the ***greatest common factor*** of $8x^2 + 12x$, since 4 would *also* divide both coefficients evenly. Are there any other factors larger than 4 that will divide **both** 8 and 12 evenly? The only other larger factor of 8 is 8 itself, which would **not** divide 12 **evenly.** Therefore 4 is the ***greatest common factor***

of both 8 and 12. So we would ***factor out*** [divide out] a 4 : $4\left(\dfrac{8x^2}{4}+\dfrac{12x}{4}\right)$

However, we have **not** factored the terms ***completely***! There is still

an x that can divide both terms evenly, and can be **factored out**: $4x\left(\dfrac{8x^2}{4x}+\dfrac{12x}{4x}\right)$

We now do the division inside the parenthesis: $4x(2x+3)$

[solution]

Factoring [dividing by a common factor] is sometimes difficult to visualize, especially when variables are involved, so it is best to **always reverse the process** [by distributing] to **check** if you've done the

factoring ***correctly***. In this case, we would **distribute**: $4x(2x+3)$

resulting in: $8x^2 + 12x$

This result is what we were asked to factor, so we have **factored *correctly*!**

Caution! If you **didn't factor _completely_**, this "check" **will _not_ detect the error**.

Let's look at this last example: If we only factored out $2x$ from $8x^2 + 12x$ the result would be:

$$2x\left(\frac{8x^2}{2x} + \frac{12x}{2x}\right)$$

After doing the division inside the parenthesis: $2x(4x+6)$

The "check" [distributing $2x(4x+6)$] results in: $8x^2 + 12x$

This is, therefore, a **"correct" factorization** [since we got back what we started with]. However, we have **_not_** factored **_completely_.** This is because $2x$ is **not** the **_greatest_** common factor of $8x^2 + 12x$. As seen earlier, $4x$ is a common factor that is **larger!**

The way we test to see if we have **_factored completely_** is to **look at what remains in the parenthesis.** These terms must **not** contain **any additional common factors.**

[Note: 1 is not considered since it is a common factor of *all* terms . Factoring out 1 doesn't change anything, since it is the identity element for division and multiplication!]

In the case of factoring out $2x$ from $8x^2 + 12x$, the result was: $2x(4x+6)$

When we examine **what remains in the parenthesis**, we see that there is **still a common factor of** 2: $[(4x+6) \Rightarrow 2(2x+3)]$ which indicates that we have **not _factored completely._** In other words, $(4x+6)$ is **not _prime_.** So we see that $2x(4x+6)$ can be **further** factored to: $2x \cdot 2(2x+3)$ which results in: $4x(2x+3)$

Now, the parenthesis, $(2x+3)$, has **no common factors** [no single factor that will divide both terms evenly other than 1] and is therefore **prime**, so the expression has been **factored _completely_!**

A common *error* in factoring:

Example 3.   **Factor:** $15b^2 + 5b$

Using the greatest common factor, $5b$, results in: $5b\left(\dfrac{15b^2}{5b} + \dfrac{5b}{5b}\right)$

Simplifying the parenthesis **_does not_** result in: $5b(3b)$

Dividing the second term in the parenthesis , $5b$, by itself results in 1.
[Note: 1 only "disappears" when it is *multiplying* another number. When it stands alone it does **_not_** disappear.]

The **correct** factorization is: $5b(3b+1)$

This can be verified by **"checking" our solution**.

Distributing $5b(3b+1)$ results in: $15b^2 + 5b$

which is what we started with! The first [incorrect] "solution" does not **"check"**. **Distributing** $5b(3b)$ results in: $15b^2$

which is _not_ what we started with!

Let's look at a more complex example. This process will require that you have a good working knowledge of the **rules of exponents.**

Example 4.   Factor: $25a^4b^5c + 20a^3b^2 - 30a^6b$

First, we make sure that every variable has an exponent. Therefore, we will add the implied exponent to the c variable in the first term, and in the b variable third term, becoming c^1 and b^1. This does **not** change their value: $25a^4b^5c^1 + 20a^3b^2 - 30a^6b^1$

Finding the greatest common factor can be confusing in such complex cases. The best approach is to analyze the **coefficients first** and then **each** variable . It is evident that each coefficient can be divided evenly by 5, which is the greatest common factor, since 10 won't divide **_every_ coefficient** evenly [It would divide 20 and 30 evenly but not 25]. Therefore, 5 i**s the greatest common factor of the coefficients:** $5\left(\dfrac{25a^4b^5c^1}{5} + \dfrac{20a^3b^2}{5} - \dfrac{30a^6b^1}{5} \right)$

Now let's look at the a variables. There is an a in **every term** so a **is a common factor**. But what power would make it the **greatest common factor**? **It can only be as large as the smallest power of a in any term in the polynomial**. If the factor were to have a larger exponent than the term that it is factoring, it would **not divide that term evenly**. It is easier to see this if we examine numbers. For example, 3^2 can't divide 3^1 evenly because 3^2, which is equivalent to 9, is too large! **For the same reason, a^4 cannot divide a^3 evenly.** If we look at this division as a fraction, $\dfrac{a^3}{a^4}$, when simplified, it **still remains a fraction** $\left[a^{-1} \Rightarrow \frac{1}{a^1} \right]$ and thus, does not divide a^3 **evenly**. We observe that a^3 is the smallest power of a in this polynomial, and therefore, a^3 is a **greatest common factor** in each of these polynomial's terms:

$$5a^3\left(\frac{25a^4b^5c^1}{5a^3}+\frac{20a^3b^2}{5a^3}-\frac{30a^6b^1}{5a^3}\right)$$

Now let's look at the b variable. There is a b in every term so b **is a factor**. But what power should it have to make it the **greatest common factor**? It can only be as large as the **smallest power of b** in any term in the polynomial [or else it wouldn't divide all the terms evenly].

We see that b^1 is the smallest power of b in the three terms of the polynomial, so b^1 **is the greatest common factor** in each of the polynomial's terms:

$$5a^3b^1\left(\frac{25a^4b^5c^1}{5a^3b^1}+\frac{20a^3b^2}{5a^3b^1}-\frac{30a^6b^1}{5a^3b^1}\right)$$

Now let's look at the c variable. There is **not** a c in all the terms of the polynomial, **so it <u>cannot</u> be factored out and** is **<u>not</u> part of the greatest common factor.**

We have now found **the greatest common factor:** $\quad 5a^3b^1$

which we factor out of the polynomial: $\quad 5a^3b^1\left(\frac{25a^4b^5c^1}{5a^3b^1}+\frac{20a^3b^2}{5a^3b^1}-\frac{30a^6b^1}{5a^3b^1}\right)$

Simplifying the terms in the parenthesis results in: $\quad 5a^3b^1\left(5a^1b^4c^1+4b^1-6a^3\right)$

Notice that $\dfrac{a^3}{a^3}$ in the second term and $\dfrac{b^1}{b^1}$ in the third term have "dropped out". That is because they are **identity elements.** Being equivalent to a 1 **multiplier**, they have **no effect** on the terms. [multiplying by 1 does not change a term's value] . Therefore they **"disappear"**.

The factorization is written more conventionally as: $\quad 5a^3b\left(5ab^4c+4b-6a^3\right)$

How do we check to see if we have factored ***completely*** and ***correctly?*** First, we notice that there are no terms in the parenthesis $\left(5ab^4c+4b-6a^3\right)$ that have common factors in **all** of the terms. If we look at the coefficients, 5, 4, and 6, there is no number [other than 1] that will divide **all of them** evenly. Also, there is not an a in every term, a b in every term, or a c in every term. Therefore, we have ***factored completely.***

We now check to see if we have ***factored correctly***
by **distributing:** $5a^3b^1\left(5a^1b^4c^1+4b^1-6a^3\right)$

Multiplying every term in the parenthesis by $5a^3b^1$ results in: $25a^4b^5c+20a^3b^2-30a^6b$

**This is the same as the original polynomial, so we have
factored correctly!**

Factoring by Grouping

This factoring technique is shown by the following examples:

Example 5.   **Factor:** $x(x+4)-5(x+4)$

Each of the two groups has a common **binomial factor** of $(x+4)$.

We can factor out $(x+4)$ by dividing each group by this factor: $(x+4)\left(\dfrac{x\cancel{(x+4)}}{\cancel{(x+4)}}+\dfrac{-5\cancel{(x+4)}}{\cancel{(x+4)}}\right)$

Can $\dfrac{(x+4)}{(x+4)}$ be canceled? **Yes**, because it is **a factor** in each term

and is also the **identity element** for multiplication.

[Any fraction with a numerator that is identical to the denominator has a value of "1".]

After the "cancellation", the resulting binomials are: $(x+4)\left(\dfrac{x}{1}-\dfrac{5}{1}\right)$

Or just: $(x+4)(x-5)$

Example 6.   **Factor:** $y^2+6y-xy-6x$

Upon careful inspection, we see that there is not a common factor in *every* term. However, we notice that the first two terms have a common factor of y and the last two terms have a common factor of x. These two groups of factors are connected by a subtraction symbol. To avoid future confusion, we will change the subtractions to addition

[and give the $-xy$ term a coefficient of -1]: $y^2+6y+(-1xy)+(-6x)$

Now, we factor out the y from the first two terms

and $-1x$ from the second two terms: $y\left(\dfrac{y^2}{y}+\dfrac{6y}{y}\right)+(-1x)\left(\dfrac{-1xy}{-1x}+\dfrac{-6x}{-1x}\right)$

Doing the division results in: $y(y+6)+(-1x)(y+6)$

Now we see that there is a **common binomial factor** in both groups which is $(y+6)$. Factoring out $(y+6)$ from each

group results in: $(y+6)\left[\dfrac{y\cancel{(y+6)}}{\cancel{(y+6)}}+\dfrac{-1x\cancel{(y+6)}}{\cancel{(y+6)}}\right]$

After "canceling" the identity elements $\dfrac{(y+6)}{(y+6)}$ from

both terms in the parenthesis: $(y+6)(y+(-1x))$

Conventionally, we change addition back to subtraction: $(y+6)(y-1x)$

and eliminate the 1 coefficient: $(y+6)(y-x)$

[Solution]

We can check our solution by distributing back **[in the case FOILing]** the solution to see if it returns with the polynomial that we started with:

FOILing $(y+6)(y-x):$ **F**irst terms: $y\cdot y=\quad y^2$

Outside Terms: $-x\cdot y=\quad -xy$

Inside Terms: $6\cdot y=\quad 6y$

Last Terms: $6(-x)=\quad -6x$

Adding the terms together gives us: $y^2-xy+6y-6x$

Changing the subtraction operations to addition: $y^2+(-xy)+6y+(-6x)$

Re-arranging the terms using the commutative property: $y^2+6y+(-xy)+(-6x)$

Changing addition back to subtraction: $y^2+6y-xy-6x$

which is the polynomial we started with!

Concept Homework

As an assessment of your understanding of the concepts set forth in this section, answer the following questions. If necessary, review the material to help you arrive at the correct conclusions. If a statement is false, correct the wording to make it true.

True or False:

1. **One technique for factoring an algebraic expression is finding the greatest factor that will divide every term in the original polynomial evenly. (Give an example.)**

2. **We can check to see if we have factored *completely* by distributing** [multiplying out] **our solution to see if it results in the original expression. (Explain your answer.)**

3. **By using the distribution process, we can check to see if we have factored *correctly*.**

4. **If we have factored completely, what remains in the parenthesis might still have a common factor. (Explain your answer.)**

5. **Factoring $15x^3 + 6x^2 + 3x$ results in $3x(5x^2 + 2x)$**

6. **The following polynomial has been factored completely: $3x\left(2x^3 + 3x^2 + 5x\right)$ (Explain your answer.)**

7. **$6x^2$ can be divided evenly by $12x^3$. (Explain your answer.)**

8. **$12x^3$ can be divided evenly by $6x^3$**

9. **$\dfrac{4x^2 y^3 - 3}{4x^2 y^3 - 3}$ is the identity element for multiplication. (Explain your answer.)**

10. **The greatest common factor of $24x^4 y^2 z + 6x^3 yz^2 - 12x^2 y^5 z^5$ is $6x^2 yz$**

11. **$y\left(z - 6\right) - 2x\left(z - 6\right)$ can be factored into $\left(y - 2x\right)\left(z - 6\right)$.**

Solutions: 1. True; 2. False; 3. True; 4. False; 5. False; 6. False; 7. False; 8. True; 9. True; 10. True; 11. True

Exercises

Factor Completely:

1. $18x - 27$ **2.** $3bx + 3b$ **3.** $4xy - 8x^2 y$ **4.** $4x^2 - 8x^3 + 12x^2$

[Hint: combine like terms before factoring.]

5. $27x^5 y - 9x^3 y^2 + 36x^4 y^3$ **6.** $2x^3 y^2 + 4x^2 y^3 - x^2 y^2$

Factor by Grouping:

7. $4(x-1) + x(x-1)$ **8.** $(a-b)c + (a-b)d$ **9.** $8r + 8s - kr - ks$

10. $6x^2 u - 3x^2 v + 2uv - v^2$ **11.** $ax^3 + bx^3 + 2ax^2 y + 2bx^2 y$

[Hint: There is a greatest common factor in all the terms!]

Solutions: **1.** $9(2x-3)$; **2.** $3b(x+1)$; **3.** $4xy(1-2x)$; **4.** $8x^2(2-x)$ or $-8x^2(x-2)$; **5.** $9x^3 y(3x^2 - y + 4xy^2)$; **6.** $x^2 y^2(2x + 4y - 1)$; **7.** $(x-1)(4+x)$; **8.** $(a-b)(c+d)$; **9.** $(r+s)(8-k)$; **10.** $(2u-v)(3x^2 + v)$; **11.** $x^2(a+b)(x+2y)$.

Topic 20 - Factoring Polynomials – Part II

Factoring Trinomials in the Form $ax^2 + bx + c$

Here we are attempting to factor a trinomial with no common **monomial** factors. As we will see in the ensuing discussion, some of these trinomials can be factored into **binomial** factors.

$ax^2 + bx + c$ is the **general trinomial form** where $"a"$ represents the coefficient of the $"x^2"$ term, $"b"$ represents the coefficient of the " x " term and $"c"$ represents some constant. Recall that the **coefficient** of x^2 is **assumed** to be $"1"$ if there is no other coefficient multiplying x^2. We have previously seen that **distributing two binomials** often results in a **trinomial**. Since factoring is the reverse of distributing, these trinomials can then be factored into two binomials!

Example 1.   **Distribute** [using the FOIL method]: $(x+2)(x+3)$

            Results in: $x^2 + 3x + 2x + 6$

            Adding like terms results in: $x^2 + 5x + 6$

The solution takes on the general trinomial form above, with the " a " coefficient assumed to be 1; 5 being the **coefficient of the** x **term** [the " b " in the general form, $ax^2 + bx + c$]; and 6 representing the **constant** [" c " in the general form, $ax^2 + bx + c$]. If

     we were asked to **factor**: $x^2 + 5x + 6$

     the result would ***have to be***: $(x+2)(x+3)$

     since **factoring** is the **reverse of distributing**.

How could we arrive at a correct factorization if we hadn't known it in advance?

     Example 2.   **Factor** : $a^2 + 7a + 12$

Strategy:
   If the trinomial is factorable, it will most often factor into two binomials.

1. Set up the parentheses for which we will eventually fill in the correct

binomial factors: $(\quad)(\quad)$

2. Fill in the **First** terms of each binomial [the "First" when we are FOILing].

Since the first term in the trinomial $[a^2 + 7a + 12]$ is a^2, the **First** terms of each binomial factor must be a. During this process, we always want to check our accuracy by **FOIL**ing our binomial factors to see if we are getting back the trinomial with which we started. Since $a^1 \cdot a^1 = a^2$, we know that our **First** terms in the binomials are: $(a \quad)(a \quad)$

3. Now we must decide on the **signs** [either **positive** or **negative**] for the **second terms** of each binomial factor. Since the last term of the trinomial, 12, in $a^2 + 7a + 12$, is **positive**, the **Last** terms in the binomial factors must **both** be positive or **both** be negative.
[**Remember: The rules for multiplying signed numbers is "positive** $\times$ **positive = positive" or "negative** $\times$ **negative = positive"]**

The **middle term,** $7a$, of the trinomial, $a^2 + 7a + 12$, is **positive.** Therefore, the addition of the **O**utside and **I**nside multiplications in the **FOIL** procedure must result in a **positive** number. This could **only** happen if the **Last** terms in each binomial were **positive**. So, the binomials should look like: $(a + \quad)(a + \quad)$

4. We now must decide on the last two **terms** of the binomials. The last term of the trinomial, $a^2 + 7a + 12$, is 12. The "**L**" in **FOIL** is the result of multiplying the last terms of the binomials. Therefore we are now looking for two factors of 12 that when **added** together, must give us the **coefficient** of the middle term which is **7**. This is the "**O+I**" part of **FOIL**! We create a table of such possible factors to analyze this situation:

possible factors of 12	which sum to **7**
$12 \cdot 1$	$12 + 1 = 13$
$6 \cdot 2$	$6 + 2 = 8$
$4 \cdot 3$	$** 4 + 3 = 7 **$

We see that 4 and 3 are the factors that we need, since $3a$ [the O in FOIL] added to $4a$ [the I in FOIL] will give us the correct middle term in $a^2 + 7a + 12$. So the correct factorization is: $(a + 4)(a + 3)$

[solution]

You should FOIL your solution to check that it gives you the trinomial you were asked to factor.

$$\text{If we FOIL } (a+4)(a+3) \Rightarrow a^2 + 3a + 4a + 12$$

$$\text{Combining like terms:} \qquad a^2 + 7a + 12$$

This is the original trinomial we were asked to factor!

Example 3.   Factor: $x^2 - 14x + 24$

<u>Strategy:</u>

1. Set up the parentheses for which we will eventually fill in the correct binomial factors: $(\quad)(\quad)$

2. Fill in the **First** terms of each binomial: $(x\quad)(x\quad)$

[If you are unsure of why the first terms need to be this way, go back to the discussion in Example 2.]

3. Now we must decide on the **signs** [either **positive** or **negative**] for the **second term** of each binomial. Since the **last term** of the trinomial, $x^2 - 14x + 24$, is **positive**, the Last terms in the binomials must <u>**both**</u> be **either** positive or negative as per the rules of multiplying signed numbers.

Since the **middle term** of, $x^2 - 14x + 24$, is $-14x$, is **negative,** the addition of the **O**utside and **I**nside multiplications in the **FOIL** procedure must result in a **negative** number. Since the trinomial has both a **positive** last term [24], and **negative** middle term [-14x], this could only happen if **both** the Last terms of the binomial factors that we are looking for were **negative.** Therefore, our binomial factors will look like: $(x-\quad)(x-\quad)$

4. We now must decide on the last two **terms** of the binomials. We are now looking for two factors of the last term in $x^2 - 14x + 24$, which is 24, such that when **added** together, must give us as the **coefficient** of the middle term, -14. This is the "**O+I**" part of **FOIL**. We create a table of such factors to analyze this situation:

possible factors of 24	which sum to -14
$(-24)(-1)$	$-24-1=-25$
$(-12)(-2)$	$**-12-2=-14**$

We see that -12 and -2 are the factors that we need, since $-2x$ [the **O** in **FOIL**] added to $-12x$ [the **I** in **FOIL**] will give us the correct middle term in $x^2-14x+24$. So the correct factorization is:

$$(x-12)(x-2)$$

[solution]

We check to make sure by **FOIL**ing: $(x-12)(x-2)$

Multiplying **F**irst Terms, **O**utside, **I**nside and **L**ast terms: $x^2-2x-12x+24$

Combining like terms: $x^2-14x+24$

Which is what we started with. Therefore, our factoring is correct!

Example 4 .   **Factor:** $x^2+4x-32$

Strategy:

1.   Set up the parentheses for which we will eventually fill in the correct binomials factors: $(\quad)(\quad)$

2.   Fill in the **F**irst terms of each binomial: $(x\quad)(x\quad)$

3.   Now we must decide on the **signs** [either **positive** or **negative**] for each binomial factor. Since the last term, -32 , of the trinomial, $x^2+4x-32$, is **negative**, the signs of the **L**ast terms in the binomial factors must be **opposite**: one **negative** and one **positive**. [Note: the rules for multiplying opposite signed numbers is: positive $\times$ negative is *always negative*!]

At this point then, the binomials should look like: $(x+\ )(x-\ )$

4.   We now must decide on the last two **terms** of the binomials. We are now looking for two factors of the last term of $x^2+4x-32$ which is -32 , when **added** together, must give us the coefficient of the middle term which is 4 .

We create a table of such factors to analyze this situation:

Possible factors of -32	Which sum to 4
$(32)(-1)$	$32+(-1)=31$
$(16)(-2)$	$16+(-2)=14$
$(8)(-4)$	$**8+(-4)=4**$

Notice in the table that the larger of the two factors must be positive so that the sum of the factors [equivalent to the middle term in the trinomial] is also positive. We see that 8 and -4 are the factors that we need, since $-4x$ [the O in FOIL] added to $8x$ [the I in FOIL] will give us the correct middle term in $x^2+4x-32$. So the correct factorization is: $(x+8)(x-4)$

[Solution]

We check to make sure by **FOIL**ing: $(x+8)(x-4)$

Resulting in: $x^2-4x+8x-32$

Combining like terms: $x^2+4x-32$

Which is what we started with!

Up to this point, we have only been dealing with trinomials that have **no common factors. This will not always be the case!**

Example 5.   Factor: $5x^3+15x^2+10x$

This trinomial is **not** in the form that we have been previously studying: $\left[x^2+bx+c\right]$. The first term's coefficient is **not** 1 and is **raised to the third power!**

However, on close examination, *we see that there is a common factor in all three terms of this trinomial that can be factored out.* It is $5x$. Factoring, we get: $5x\left(\dfrac{5x^3}{5x^1}+\dfrac{15x^2}{5x^1}+\dfrac{10x}{5x^1}\right)$

Which results in: $5x\left(x^2+3x+2\right)$

We are **not finished yet** since **the remaining trinomial,**
$$x^2 + 3x + 2, \textbf{ can be factored further:}$$
$$5x\left(x+\right)\left(x+\right)$$

We are now looking for two factors of the last term, 2, of the trinomial in the parenthesis $\left(x^2 + 3x + 2\right)$. When these factors are added together they must give us 3, the coefficient of the middle term. We create a table of such factors to analyze this situation:

Possible factors of **2**	Which sum to **3**
$2 \cdot 1$	$** 2 + 1 = 3 **$

We see that 2 and 1 are the factors that we need: $5x\left(x+2\right)\left(x+1\right)$

[Don't leave out the $5x$ along the way! It is part of the factorization.]

Let's multiply these factors to see if we have factored correctly: $5x\left(x+2\right)\left(x+1\right)$

1.   Start by multiplying the binomials together: $5x\left(x^2 + 1x + 2x + 2\right)$

2.   Combining like terms within the parenthesis: $5x\left(x^2 + 3x + 2\right)$

3. Multiplying all terms within the parenthesis by $5x$: $5x^3 + 15x^2 + 10x$

It checks! This is the trinomial that we started with!

Special Binomial Factoring

"The Difference [subtraction] of Two Perfect Squares": $a^2 - b^2$

This is a **special case** that we should be able to recognize. It occurs when factoring **the difference of two perfect squares.**

Example 6.   **Factor:** $x^2 - 25$

Even though this algebraic expression is **not a trinomial, factoring will make two binomial factors in this special case.**

Notice that x^2 and 25 are **perfect squares** $\left[x \cdot x = x^2 \text{ and } 5 \cdot 5 = 25\right]$.

Let's see **why** this binomial will factor into two simpler binomials.

<u>Strategy:</u>

1. Set up the parentheses for which we will eventually fill in the correct binomials: $(\quad)(\quad)$

2. Fill in the **First** terms of each binomial: $(x\quad)(x\quad)$

3. Now we must decide on the **signs** [either **positive** or **negative**] for the **second term** of each binomial. Since the last term of $x^2 - 25$ is **negative**, the signs of the **Last** terms in the binomial factors must be **opposites:** one **positive** and one **negative**. [Note: **positive** $\times$ **negative** is *always negative!*]

At this point then, the binomials should look like: $(x +\quad)(x -\quad)$

4. We now must decide on the last two **terms** of the binomials.

Since there is *no middle term* **in this trinomial** the **Outside** and **Inside** multiplications **must add to zero!** [They must be opposites!] We are now looking for two factors of the last term of $x^2 - 25$ that, will add to 0.

We create a table of such factors to analyze this situation:

Possible factors of -25	Which sum to 0
$(25)(-1)$	$25 + (-1) = 24$
$(5)(-5)$	$**\, 5 + (-5) = 0 \, **$

The factors 5 and -5 are the ones that we need: $(x + 5)(x - 5)$

[Solution]

Checking, we see that FOILing these factors: $(x + 5)(x - 5)$

gives us: $x^2 + 5x - 5x - 25$

Combining like terms eliminates a "middle term": $x^2 - 25$

This binomial is what we started with!

Note: The solution's binomial factors are called **<u>conjugates!</u>**

[The binomials are identical except one has addition and the other subtraction: $(x+5)(x-5)$]

Conjugates are special because, when we FOIL them, there is never a middle term! This special factoring case, "factoring the difference [subtraction] between perfect squares", always results in conjugates.

Here is a more complex example of "factoring the difference between perfect squares":

Example 7.   **Factor:** $x^4 - 16$

   This binomial is indeed the **difference between perfect squares**
   since both x^4, $\left[x^4 = x^2 \cdot x^2 \right]$, and 16, $[16 = 4 \cdot 4]$, are perfect squares

connected by subtraction! Therefore, they will factor into: $\left(x^2 + 4 \right)\left(x^2 - 4 \right)$

[If you FOIL these binomial factors they will result in $x^4 - 16$]
As expected, the results of the factoring are **conjugates. But we
are not finished!** Notice that the **second binomial factor** $\left(x^2 - 4 \right)$

is *also* the **difference between perfect squares.**

[both x^2 and 4 are perfect squares connected by subtraction.]

Factoring $\left(x^2 - 4 \right)$ results in: $(x+2)(x-2)$. So now the ***complete***

factoring is: $\left(x^2 + 4 \right)\left(x + 2 \right)\left(x - 2 \right)$

[solution]

Now, $x^4 - 16$ **is factored *completely*** .

[Note that $\left(x^2 + 4 \right)$, the first factor, is *not* factorable! It is the *addition* of perfect squares, not the difference of

perfect squares. You might think that $\left(x^2 + 4 \right)$ factors into $(x+2)(x+2)$. It does *not*. $x^2 + 4x + 4 = (x+2)(x+2)$.]

Factoring Trinomials Whose Leading Coefficient is *Not* "1"

Up until now, our study of factoring trinomials were limited to the general form $x^2 + bx + c$ which implies that the coefficient of the x^2 term is 1. Suppose the x^2 term had a **larger coefficient**. The general form would then become $ax^2 + bx + c$, where $"a"$ represents the coefficient of x^2 [called the "leading coefficient"] which is not necessarily "1", $"b"$ represents the coefficient of the middle term, and $"c"$ represents the constant.

To factor these types of trinomials, we will use the "$a \cdot c$ method" along with **factoring by grouping** which we discussed earlier. Here is how it works:

Example 8.   **Factor:** $2x^2 + 5x - 3$

Since the leading coefficient is not 1, we will employ the "$a \cdot c$ **method**". With this trinomial, multiplying "a" [the leading coefficient of the x^2 term] by "c" [the constant] would be $(2)(-3) = -6$.

The middle term, $5x$, with a coefficient of 5, is the "b" coefficient. We need to **replace** this term, $5x$, with **two terms** *whose sum is equivalent to* $5x$. These terms <u>also</u> have to be factors of -6 [the result of multiplying $a \cdot c$].

We create a table of such factors to analyze this situation:

possible factors of $a \cdot c$ = -6	that sum to "b" or 5
$(-6)(1)$	$-6 + 1 = -5$
$(6)(-1)$	$**6 + (-1)**$

Notice that the larger factor must be positive in order to end up with a positive middle term. The factors 6 and -1 give us what we need. We **replace** the middle term of $2x^2 + 5x - 3$, which is $5x$, with its **equivalent: two terms that sum to** $5x$. The terms that we will use are $6x$ and $-1x$. We have not altered the value of the original trinomial since $\left[6x + (-1x) = 5x\right]$. Replacing $5x$ with its equivalent, $6x + (-5x)$

gives us the **equivalent** polynomial: $2x^2 + 6x + (-1x) - 3$

We now employ the "factoring by grouping" method examined in the previous factoring topic. The first two terms can be factored as well as the last two terms: $2x\left(\dfrac{2x^2}{2x^1} + \dfrac{6x^1}{2x^1}\right) + (-1)\left(\dfrac{-1x}{-1} + \dfrac{-3}{-1}\right)$

The division within the parenthesis results in: $2x(x + 3) - 1(x + 3)$

Now we see that there is a common **binomial** factor in both groups which is $(x + 3)$. Factoring out $(x + 3)$ from each group results in: $(x + 3)(2x - 1)$

[Solution]

[**Check** to see if this is right by FOILing $(x+3)(2x-1)$ to see if it results in $2x^2+5x-3$]

Example 9. - Application

A rectangle has an **Area** of $2x^2+11x+5$. In terms of x, **what are its length and width** ?

Since in a rectangle $A = l \cdot w$, we can determine the length and width if we can **factor** $2x^2+11x+5$ [equivalent to the Area] into two binomials which will then represent l and w.

So we must _factor_: $2x^2+11x+5$

Since the leading coefficient is not 1, we will employ the "$a \cdot c$ **method** ".With this trinomial, multiplying "a" the leading coefficient of the x^2 term] by "c" [the constant] would be $(2)(5)=10$.

The middle term, $11x$, with a coefficient of 11, is the "b" coefficient. We need to **replace** this term, $11x$, with **two erms _whose sum is equivalent to_ $11x$**. These terms **also** have to be factors of 10 [the result of multiplying $a \cdot c$].

We create a table of such factors to analyze this situation:

possible factors of $a \cdot c = 10$	that sum to "b" or 11
$(10)(1)$	$**10+1=11**$

The factors 10 and 1 give us what we need. We **replace** the middle term of $2x^2+11x+5$ which is $11x$, with its **equivalent - two terms that sum to** $11x$. The terms that we will use are $10x$ and $1x$. We have not altered the value of the original trinomial since $10x+1x=11x$. Replacing $11x$ with its equivalent, $10x+1x$, gives us the equivalent

polynomial: $2x^2+10x+1x+5$

We now employ the "factoring by grouping" method
The first two terms can be factored as well as the

last two terms: $2x\left(\dfrac{2x^2}{2x^1}+\dfrac{10x^1}{2x^1}\right)+1\left(\dfrac{1x}{1}+\dfrac{5}{1}\right)$

The division within the parenthesis results in: $2x(x+5)+1(x+5)$

Now we see that there is a common **binomial**
factor in both groups which is $(x+5)$. Factoring out

$(x+5)$ from each group results in: $(x+5)(2x+1)$

Now, going back to the original problem:
$A=2x^2+11x+5$ **[the area of the rectangle]**

In the factored version of $2x^2+11x+5$, $A=(x+5)(2x+1)$.

But $A=\ l\ \cdot\ w$

Therefore, $l=x+5$ and $w=2x+1$ [or vice versa]
[Solution]

Concept Homework

As an assessment of your understanding of the concepts set forth in this section, answer the following questions. If necessary, review the material to help you arrive at the correct conclusions. With "true or false" statements, if the statement is true, make up an example that supports the statement without using examples given in the material. If a statement is false, correct the wording to make it true. Then make up an example that supports the statement without using examples given in the material.

True or False

1. When factoring a trinomial into two binomial factors, the "outside-inside" [the OI of FOIL] multiplications must add up to the middle term of the trinomial.

2. Factoring the *difference* [subtraction] between two perfect squares will result in binomial factors that are *identical*.

3. Factoring $x^2 + 2x - 24$ will result in binomials where the last terms will have the same signs.

4. When we multiply conjugates, the result will always be a trinomial.

5. When we expand $(a-b)^2$, there will be no "middle term".

6. If we factor $x^2 + 8x - 12$, the result will be two binomials both connected by addition.

7. Conjugates are identical binomial factors except that one has the addition operation and the other has the subtraction operation. When expanded, they will have no middle term.

8. $x^4 - 81$ factors into $(x^2 + 9)(x^2 - 9)$ but is not factored completely.

9. $(x^2 + y^2)$ factors into $(x+y)(x+y)$.

10. $(m^2 - n^2)$ factors into $(m+n)(m-n)$.

Solutions: **1.** True; **2.** False; **3.** False; **4.** False; **5.** False; **6.** False; **7.** True; **8.** True; **9.** False [not factorable. Therefore the polynomial is "prime".]; **10.** True.

Exercises

Completely factor the following trinomials:

1. $x^2 + 5x + 6$　　**2.** $x^2 - 3x - 10$　　**3.** $a^2 + 2a - 35$　　**4.** $x^2 + 4ax + 4a^2$

[Hint: the template is: $(x \quad ?a)(x \quad ?a)$]

5. $-x^2 + 2x + 15$　　　　**6.** $-3ax^2 - 6ax + 9a$　　　　**7.** $25x^2 - 36$

[Hint: First factor out -1]　　[Hint: First factor out the greatest common factor]

8. $x^3 - xy^2$　　　　**9.** $y^4 - 16$　　　　**10.** $4x^2 + 8x + 3$

[Hint: Make sure you factor completely.]　　[Use the $a \cdot c$ method]

11. $3a^2 - 4a - 4$　　　　**12.** $25x^2 + 16y^2$

13. If a rectangle has an area of $2x^2 + 5x - 12$, in terms of x , what is its length and width [in terms of x] ?

Solutions: **1.** $(x+2)(x+3)$; **2.** $(x-5)(x+2)$; **3.** $(a+7)(a-5)$; **4.** $(x+2a)(x+2a)$ or $(x+2a)^2$; **5.** $-(x-5)(x+3)$;

　　6. $-3a(x+3)(x-1)$; **7.** $(5x+6)(5x-6)$; **8.** $x(x+y)(x-y)$; **9.** $(y^2+4)(y+2)(y-2)$; **10.** $(2x+3)(2x+1)$;

　　11. $(a-2)(3a+2)$; **12.** Can't be factored. Therefore the polynomial is "prime"; **13.** Length: $2x-3$; width: $x+4$ (or vice versa).

Topic 21 – Using Factoring to Solve Quadratic Equations

You might be wondering why mathematicians developed algebraic factoring techniques. It was done to help solve equations! In this topic we will discuss how factoring can help solve **quadratic equations**.

Using Factoring to Solve Quadratic Equations

Quadratic equations are 2nd degree equations in one variable with the right side set equal to 0. 2nd degree equations have the variable raised to the 2nd power and no higher. Such equations are in the general form :

$$Ax^2 + Bx + C = 0$$

where A is a non-zero number **[coefficient of the x^2 term]**, B represents the coefficient of the x term, and C represents some constant. Therefore, the only actual variable in the quadratic equation are the $x's$ in the preceding equation.

Example 1. Solve for x : $x^2 - 3x + 2 = 0$

In this form, it is not immediately evident as to how to start finding values for x that will make the equation true other than trial and error, a very inefficient way of solving equations. However, if we **factored the left side of the equation**, this problem is simplified.

Using techniques from the previous topics, we know that $x^2 - 3x + 2$ can be factored into $(x-2)(x-1)$, **so we can *rewrite* the equation:** $(x-2)(x-1) = 0$

How does this help us solve for x ?

It helps to think of the **entire binomial,** $(x-2)$, as being equivalent to some number value, say represented by **P,** that is multiplying the **binomial,** $(x-1)$, equivalent to some other number value, represented by **Q**. Then, the equation, $(x-2)(x-1) = 0$, becomes: $(P) \cdot (Q) = 0$

Now let's experiment with this simple equation, $P \cdot Q = 0$. There are an infinite number of values that we can either assign to P or Q to find the other value that will make the equation true. Here are just a few:

Let $P = 3$: in $P \cdot Q = 0$
Substituting 3 for P: $3 \cdot Q = 0$

In order for this equation to be true, Q would have to be equivalent to 0.

Suppose we let $P = -5$ in $P \cdot Q = 0$
Substituting −5 for P: $-5 \cdot Q = 0$

In order for this equation to be true, Q would still have to be equivalent to 0.
I think you will agree that *any value* assigned to **P** would not change the fact that
Q would **have to be equivalent to 0** to make the equation $P \cdot Q = 0$ true! [Any number
multiplied by 0 will always result in 0.]

Now, suppose we had assigned values to Q?

$$\text{Let } Q = 4: \text{ in} \qquad P \cdot Q = 0$$
$$\text{Substituting } 4 \text{ for Q:} \qquad P \cdot 4 = 0$$

In order for this equation to be true, P would have to be equivalent to 0.

$$\text{Suppose we let } Q = -7 \text{ in} \qquad (P) \cdot (Q) = 0$$
$$\text{Substituting } -7 \text{ for Q:} \qquad (P) \cdot (-7) = 0$$

In order for this equation to be true, P would still have to be equivalent to 0.

I think you will agree that *any value* assigned to **Q** would not change the fact that
P would **have to be equivalent to 0** to make the equation $(P) \cdot (Q) = 0$ true!

So, one of the following statements must be true for: $\qquad P \cdot Q = 0$

$$\textbf{Either } P = 0, \text{ since the equation would then be:} \qquad 0 \cdot Q = 0,$$
$$\textbf{or } Q = 0, \text{ in which case the equation would then be:} \qquad P \cdot 0 = 0.$$
[It is also possible that *both* P and Q are equal to 0.]

**The only way to multiply two factors so that they result in 0 is when at
least one of the factors has a value of 0.** [*Only* when a number is multiplied by 0, will
their product be 0.]

$$\text{\textbf{Now}, going back to the \textbf{earlier} equation:} \qquad x^2 - 3x + 2 = 0$$
$$\text{which was factored into:} \qquad (x - 2)(x - 1) = 0$$
$$\text{Let's think of this equation as if it were} \qquad P \quad \cdot \quad Q \quad = \quad 0$$

$(x - 2)$ is represented by "**P**" and $(x - 1)$ is represented by "**Q**".

This would mean the either $(x - 2)$ has to have a value of 0

or $(x - 1)$ has a value of 0.

$$\textbf{Suppose:} \qquad \textbf{(1)} \qquad x - 2 = 0$$
[the parenthesis is no longer necessary]

$$\text{adding } 2 \text{ to each side of the equation:} \qquad \underline{ 2 \quad 2}$$
$$\text{Then:} \qquad x \quad = 2$$
[one possible solution]

Looking at the other possibility: **(2)** $x - 1 = 0$
[the parenthesis is no longer necessary]

adding 1 to each side the equation: $\underline{\quad\quad 1 \quad 1}$

Then: $x \quad = 1$

[another possible solution]

Therefore, there are **two** the solutions to $x^2 - 3x + 2 = 0$: $\quad x = 2$ and $x = 1$

Note: We have seen that <u>linear equations</u> in one variable usually have only <u>one</u> solution. However, quadratic equations usually have <u>two</u> solutions!

It is always a good idea to verify that the solutions really satisfy the original equation.

(1) Substituting $x = 2$ into $x^2 - 3x + 2 = 0$: $\quad 2^2 - 3(2) + 2 = 0$

Simplifying: $\quad 4 - 6 + 2 = 0$

Results in: $\quad -2 + 2 = 0$

[which is true!]

(2) Substituting $x = 1$ into $x^2 - 3x + 2 = 0$: $\quad 1^2 - 3 \cdot 1 + 2 = 0$

Simplifying: $\quad 1 - 3 + 2 = 0$

Resulting in: $\quad -2 + 2 = 0$

[which is also true!]

So we know that our solutions are correct!

Sometimes, quadratic equations are **disguised.** By that we mean, they are not set equal to zero. When that is the case, all we need to do is **"un-disguise"** them as in the following:

Example 2. **Solve for** x: $\quad x^2 - x = 72$

We need to **set the equation to zero for factoring to be useful**. Otherwise, our $P \cdot Q = 0$ strategy would no **longer be valid**. We do this by adding -72 to both sides of the equation:

$$x^2 - 1x \quad\quad = \quad 72$$
$$\underline{\quad -72 \quad\quad -72 \quad}$$

So, we have

accomplished this: $\quad x^2 - 1x - 72 = 0$

Now our factoring strategy can be employed. Factoring the

Left side of the equation results in: $\quad (x - 9)(x + 8) = 0$

We know that one of these two binomials must equal zero for the equation to be true.

[Two factors being multiplied can only equal 0 when one of the factors has a value of 0!]

So we set <u>**each**</u> binomial equal to zero.   **(1)** $x - 9 = 0$

Adding 9 to both sides of the equation: $\underline{9 \quad 9}$

Results in: $x \quad = 9$

[one solution]

Now we set the second binomial equal to zero. **(2)** $x + 8 = 0$

Adding -8 to both sides of the equation: $\underline{-8 \quad -8}$

Results in: $x \quad = -8$

[another solution]

So our solutions to $x^2 - x = 72$ are: $x = 9$ and $x = -8$

As in the previous example, we can substitute the
solutions into the original equation: $x^2 - x = 72$

To see if they indeed make it true: If $x = 9$, then: $9^2 - 9 = 72$

Simplifying: $81 - 9 = 72$

[This is a true statement.]

If $x = -8$, then: $(-8)^2 - (-8) = 72$

Simplifying: $64 + 8 = 72$

[This is a true statement.]

Next is an example where the **quadratic is NOT a trinomial**:

Example 3.   Solve for x: $x^2 - 9 = 0$

The left side of the equation is not the **trinomial** that we normally
see in quadratic equations, but **it is factorable!** It is that **special
case** where **two perfect squares are** connected by a **subtraction
operation.** [the difference between two perfect squares!]

This kind of quadratic will factor into <u>***conjugates***</u>

that, when multiplied, **never have a middle term:** $(x + 3)(x - 3) = 0$

[If you multiply these binomials, you will see that the middle

terms add to zero, resulting in $x^2 - 9$]

We know that one of these two binomials must equal zero for
the equation to be true. So we set each binomial equal to zero:

(1) $x + 3 = 0$ or **(2)** $x - 3 = 0$

Isolating x in each equation: $\underline{-3 \quad -3}$ $\underline{3 \quad 3}$

results in: $x = -3$ or $x = 3$

[one solution]   [another solution]

{ 190 }

So our solutions to $x^2 - 9 = 0$ **are:** $x = -3$ and $x = 3$

As an exercise, substitute the solutions into the original equation, $x^2 - 9 = 0$, to see if they indeed make it true.

Here is another quadratic that is **not** a trinomial and is **not** the difference of perfect squares. However, **we can use factoring to solve this equation!**

Example 4.   **Solve for** x: $3x^2 - 27 = 0$

First, we notice **that we can factor** 3 from the binomial on the

left side: $3(x^2 - 9) = 0$

Now we can divide both sides of the equation by 3: $\dfrac{3(x^2 - 9)}{3} = \dfrac{0}{3}$

The 3's cancel out on the left side and $\dfrac{0}{3} = 0$ on the right side: $\dfrac{\cancel{3}(x^2 - 9)}{\cancel{3}} = 0$

[Note: 0 divided by any non-zero number equals 0. This is different from dividing *by* 0, which cannot be done.]

By factoring out the 3, we have created

"the difference of perfect squares": $x^2 - 9 = 0$

We already found the solutions to this equation in **Example 3**: $x = -3$ and $x = 3$

Sometimes, factoring quadratic equations **requires a different strategy** as in the following example.

Example 5.   **Solve for** x: $4x^2 - 2x = 0$

Notice that there is **no constant** in this binomial! This means that an x term **can be factored out!**

[The greatest common factor of this quadratic is $2x$.]

Therefore, we can **factor out** $2x$: $(2x)\left(\dfrac{4x^2}{2x} + \dfrac{-2x}{2x} \right) = 0$

Resulting in: $(2x)(2x - 1) = 0$

Just as in the previous examples, one of these two factors must take on the value of 0 for this equation to be true.

Suppose: **(1)** $2x = 0$

Dividing both sides by 2: $\quad \dfrac{2x}{2}=\dfrac{0}{2}$

Results in: $\quad x=0$

[Note: 0 divided by any non-zero number equals 0. Think of $\frac{0}{2}$ as zero parts of a whole that is broken into 2 equal parts. This would be zero.]

The other possibility is: **(2)** $\quad 2x-1=0$

adding 1 to both sides: $\quad \dfrac{1 \quad\; 1}{2x \;\;\; = \; 1}$

Results in:

Dividing both sides by 2 : $\quad \dfrac{2x}{2}=\dfrac{1}{2}$

Results in: $\quad x=\dfrac{1}{2}$

[second solution]

$\qquad$ **So the solutions to** $4x^2-2x=0$ **are:** $x=0$ **and** $\quad x=\dfrac{1}{2}$

Verifying our answers:

Let $x=0$ in $4x^2-2x=0 \;\Rightarrow\; 4(0)^2-2(0)=0 \;\Rightarrow\; 0-0=0$ [true statement]

Let $x=\dfrac{1}{2}$ in $4x^2-2x=0 \;\Rightarrow\; 4\left(\dfrac{1}{2}\right)^2-\dfrac{2}{1}\left(\dfrac{1}{2}\right)=0 \;\Rightarrow\; \dfrac{4}{1}\cdot\dfrac{1}{4}-\dfrac{2}{1}\cdot\dfrac{1}{2}=0 \;\Rightarrow\; 1-1=0$ [true statement]

An Application that Uses Quadratic Equations

Example 6.

The **length** of a **rectangle** is **5 inches longer** than the **width**. The **area** of the rectangle is 66 square inches. Find the rectangle's length and width.

First we ask ourselves, **"What do we know about a rectangle that is pertinent to the information given?"**

We know that $\textbf{\textit{L}}\text{ength} \cdot \textbf{\textit{W}}\text{idth} = \textbf{\textit{A}}\text{rea}$. We use this formula to write an equation by plugging in the given information. **The piece of information that we know the least about is the width**, so we will assign it the variable $\textit{W}$. **The length can be expressed in terms of** $\textit{W}$. Here is a diagram of the problem with variables assigned to the length and width:

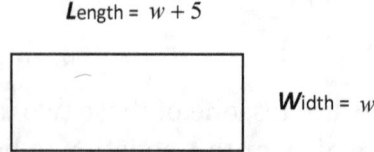

$\textbf{\textit{L}}\text{ength} = w+5$

$\textbf{\textit{W}}\text{idth} = w$

To make a quadratic equation, we can **only have one variable!**

We can accomplish this by **eliminating the other variable, _L_,**
by defining the length in terms of the width: Length $= w + 5$
[This information was given to us in the problem.]

Also given to us was the **_A_rea:**   Area $= 66$
Substituting $w + 5$ for the **_L_ength**, and using
_L_ength $\cdot$ _W_idth $= $ _A_rea results in: $(w+5) \cdot w = 66$

Or more conventionally: $w(w+5) = 66$
[We employed the commutative property of multiplication.]

Distributing results in: $w^2 + 5w = 66$
[a quadratic equation!...but not set equal to zero.]

We need to set the equation equal to zero: $w^2 + 5w \quad = \quad 66$
We accomplish this by adding -66 to both sides: $\underline{-66 \quad\quad -66}$

Results in: $w^2 + 5w - 66 = 0$
We now factor the left side of the equation: $(w+11)(w-6) = 0$

We know that **one of these two binomials must equal zero**
for the equation to be true. So we set each binomial equal to zero.

Suppose: **(1)** $w + 11 \quad = \quad 0$
Adding -11 to both sides: $\underline{-11 \quad -11}$

Results in: $w = -11$
[one "solution"]

The other possibility is: **(2)** $w - 6 \quad = \quad 0$
Adding 6 to both sides: $\underline{6 \quad\quad 6}$
Results in: $w \quad = \quad 6$
[second solution]

So our "solutions" are: $w = -11$ inches **and** $w = 6$ inches.

However, since a **width can't be negative** [distances are always ≥ 0],
we reject the -11 "solution" and keep the **_W_idth** as: **6 inches**

The **_L_ength,** $w + 5$, is $6 + 5 = $ **11 inches**

We now verify that the answer is correct :   **_L_ength $\cdot$ _W_idth $= $ _A_rea** and, indeed,
$11 \cdot 6 = 66$ **square inches**

Which is what was given as the area of the rectangle!

Concept Homework

If an answer is false, explain why or show how you would change the statement to make it true. If you are asked to "explain", sometimes it helps to give an example.

True or False:

1. The only way to multiply two factors so that they result in 0 is when one or both of the factors has a value of 1.

2. In its present format, $x^2 - 2x = 24$, factoring the left side of the equation will not help us find a solution. [If true, explain why.]

3. The equation $4x^2 - 36 = 0$ can be solved by factoring the left side of the equation even though there is no "middle term" as there are with trinomials.
   [If true, explain how we can factor when there is no middle term.]

4. The left side of a quadratic equation must be a trinomial in order to use factoring to find solutions. [If false, give an example not used in these notes.]

5. The equation $5x^2 + 10x = 0$ can be solved by a factoring method.
   [If true, what are the solutions?]

6. An equation with one variable that is to the first power usually has a unique [only one] solution that makes it true. This is also usually true of quadratic equations.

7. A quadratic equation has a variable to the second power and no higher.

9. By factoring a quadratic equation, we can sometimes create two single-variable equations whose variable is to the first power. [If true, give an example not given in the notes.]

Solutions: 1. False; 2. True; 3.True; 4. False; 5. True; 6. False; 7. True; 8. True.

Exercises

Solve the following equations:

1. $x^2 - 3x + 2 = 0$ 2. $x^2 + 4x - 21 = 0$ 3. $x^2 + 14x = 15$ 4. $2x^2 - 10x = 12$

5. $4x^2 - 8x = 0$ 6. $2x^2 + 6x - 7 = x$ 7. $6x^2 - 54 = 0$
[Hint: Get the right side to equal zero. Then use the $a \cdot c$ method to factor.]

8. The square of some integer is 7 less than 8 times the integer. Find the integer(s).
[Hint: Let x be "some integer". Then translate the sentence into a mathematical equation. The word "is" translates into "=".]

Solutions: **1.** $x = 1$ or $x = 2$; **2.** $x = -7$ or $x = 3$; **3.** $x = -15$ or $x = 1$; **4.** $x = 6$ or $x = -1$; **5.** $x = 0$ or $x = 2$;

6. $x = 1$ or $x = -\dfrac{7}{2}$; **7.** $x = 3$ or $x = -3$; **8.** The integer is either 7 or 1.

Topic 22 - Multiplication and Division with Rational Expressions (Algebraic Fractions)

Rational numbers are real numbers that can be expressed as fractions. The term "rational" in this sense is related to the meaning of "ratio"[another name for fraction] rather than the non-mathematical meaning of having the ability to reason. When variables are involved in rational numbers, the fractions are called **rational expressions**. Therefore, rational expressions can be understood as "**algebraic fractions**", or fractions with variables in the numerator, denominator, or both.

Examples: (a) $\dfrac{n^2+2n-1}{n+1}$ (b) $-\dfrac{a^2}{7bc}$ (c) $\dfrac{5}{x-2}$

We have previously discussed how to **simplify** polynomials through factoring. We will now discuss how to **multiply, divide** and **simplify rational expressions** [algebraic fractions] using the **factoring process**.

We must always keep in mind that the rational expressions can become **undefined** if the variables take on values that will make the denominator equivalent to 0 [a fraction with 0 in the denominator is undefined].

- In **example (a)** above, if n took on the value of -1, the denominator would become 0 and the fraction would become undefined.
- In **example (b)** above, if b or c [or both] took on the value of 0, the denominator would become 0 and the fraction would become undefined.
- In **example (c)** above, if x took on the value of 2, the denominator would become 0 and the fraction would become undefined.

Therefore, with any rational expression, **we must be aware of those values for the variable that would make the denominator 0 and exclude them from consideration.**

Multiplication of Rational Expressions

This is just an extension of the **simplification** process. In many cases, it will require **factoring**.

First let's review how we multiply regular fractions:

Example 1 [without variables].   Multiply: $\dfrac{6}{10} \cdot \dfrac{4}{3}$

The rule for multiplying fractions is to treat the numerators and denominators as a series of factors [numbers multiplying each other]. Many times, it is convenient to leave them as factors rather than actually doing the multiplication. In effect, we are multiplying numerators together and denominators together. Since there is **only the multiplication operation involved**, we can remove identity elements (cancel) to simplify the problem. This is done by breaking down the

original numbers into *convenient* factors to create identity elements: $\dfrac{2 \cdot 3 \cdot 4}{2 \cdot 5 \cdot 3}$

Because of the **commutative property** of multiplication we can **re-arrange**

the factors to clearly see the **identity elements**: $\dfrac{\cancel{2} \cdot \cancel{3} \cdot 4}{\cancel{2} \cdot \cancel{3} \cdot 5}$

The identity elements, are in effect multiplying the entire fraction by "1" which does not affect the fraction's value . Therefore, we can **remove them** from the fraction without changing the fraction's quantitative value [canceling].

Canceling is **really a procedure for removing identity elements:** $\dfrac{4}{5}$

[solution]

We streamline this process of "cancelling" by dividing out common factors [identity elements] in the numerator and denominator. Let's see

how this works with the same problem: $\dfrac{6}{10} \cdot \dfrac{4}{3} \Rightarrow \dfrac{\overset{2}{\cancel{6}} \cdot \overset{2}{\cancel{4}}}{\underset{5}{\cancel{10}} \cdot \underset{1}{\cancel{3}}}$

Here we have divided the 6 in the numerator and 3 in the denominator by 3, and the 4 in the numerator and 10 in the denominator by 2 resulting in: $\dfrac{2 \cdot 2}{5 \cdot 1}$

Multiplying the numerators and denominators

Gives us the same results as the previous approach: $\dfrac{4}{5}$

The same process occurs when we multiply algebraic fractions:

Through the factoring process, we are able to remove **"polynomial identity elements"** for multiplication. It is important to understand that identity elements can have variables

contained in them. For example, $\dfrac{x}{x}$, $\dfrac{x-2}{x-2}$, and $\dfrac{x^2-3x-5}{x^2-3x-5}$ are **all identity elements**

and have the value of "1" . In fact, **any fraction whose numerator and denominator are identical** are **identity elements for multiplication** and have a value of "1" .

Example 2. **Simplify the rational expression (algebraic fraction):** $\dfrac{x^2 + x - \cancel{12}}{x + 4}$

[only when $x \neq -4$ which would make the denominator zero.]

It is not immediately evident as to where to start. We **cannot** "cancel" each x term in the numerator and denominator and **cannot** divide out a 4 from the 12 as we are tempted to do

which would result in: $x^2 + 1 - 3$
[incorrect simplification]

Why not? There are **subtraction/addition operations** that exist in both the numerator and denominator. The only time we are able to "cancel" is when the algebraic fraction is made up of **factors only** [multipliers].

In order to "cancel" terms, **the numerator and denominator must be made exclusively of factors** [a series of expressions that are multiplying each other].

For example, if the original fraction had **only multiplication operations:** $\dfrac{x^2 \cdot x \cdot (-12)}{x \cdot 4}$

[only when $x \neq 0$]

Then we **could** eliminate an x term in the numerator and denominator

and divide both -12 and 4 by 4, resulting in: $\dfrac{x^2 \cdot 1 \cdot (-3)}{1}$

or more conventionally: $-3x^2$

Because of **the addition and subtraction operations** in $\dfrac{x^2 + x - 12}{x + 4}$, we **cannot** "cancel" terms since they are **not factors**.

So what **can** we do to simplify? $\dfrac{x^2 + 1x - 12}{x + 4}$

[The "hidden" coefficient "1" is added to the x term in the numerator to remind us that it is *there*, even when it is not written. After all, multiplying by the identity element, 1, doesn't change the value of x.]

The answer lies in **factoring**. The numerator, $x^2 + 1x - 12$, can be factored into $(x+4)(x-3)$. So we can express the fraction above as: $\dfrac{(x+4)(x-3)}{(x+4)}$

[Putting a parenthesis around $x + 4$ in the denominator doesn't change its value.]

We see that **now** the numerator has two **binomial factors** [multipliers], one of which is **identical to the denominator** $(x+4)$. **These <u>can</u> be "canceled" out, since they are actually a "polynomial identity element" for multiplication:** $\dfrac{(x+4)}{(x+4)} = 1$. Even though there are addition operations *within* them, **they are both identical *factors*. We call them binomial factors.**

[There are no addition/subtraction operations *between* the parentheses in the numerator or denominator. Therefore, they can be treated as factors]

If we cancel the binomial factors $(x+4)$:
$$\frac{\cancel{(x+4)}\,(x-3)}{\cancel{(x+4)}}$$

The canceling results in a **simplified version of the fraction**:
$$\frac{(x-3)}{1}$$

Or just:
$$x-3$$

We will now discuss the **problem** with this simplification. **The value of** x **is supposed to represent <u>any</u> real number value.** Our simplification process resulted in $\dfrac{x^2+x-12}{x+4}$ being **equivalent to** $x-3$ **for <u>any</u> value assigned to** x. Now, suppose the value of x happened to be -4. f we **evaluated** the

original fraction:
$$\frac{x^2+x-12}{x+4}$$

with $x=-4$ [**substituting for** x **throughout the equation**]:
$$\frac{(-4)^2-4-12}{-4+4}$$

This would simplify to:
$$\frac{16-4-12}{-4+4}$$

Add up the numerators and denominators would result in:
$$\frac{0}{0}$$

and the denominator would become zero! We **know** that the **denominator** of a fraction **cannot be zero**, or the fraction is **undefined**!

Therefore, when we are asked to simplify $\dfrac{x^2+x-12}{x+4}$, x can be any real number **except** a number that makes the denominator have a quantitative value of 0. In this case, **that number is** -4. So, we stipulate $x \neq -4$. **In a rational expression that has a variable(s) in the denominator**, there **will** be some value(s) that could be assigned to x that would give the denominator a value of 0. This value(s) must be explicitly excluded. This is the reason, in this case, for stating "$x \neq -4$".

How can we find those value(s)?

Simply set the denominator equal to 0 and solve the equation!

In this case: $\qquad x + 4 = 0$

Adding -4 to both sides: $\qquad \dfrac{-4 \quad -4}{}$

Results in: $\qquad x \quad = -4$

Example 3.

For what value of x is the following fraction undefined? $\qquad \dfrac{2x^2 + 3x - 5}{2x - 6}$

When the **whole** denominator [not just the *x* value] of this fraction is equal to 0, the whole fraction is undefined. This will happen when the whole denominator, $2x - 6$, is equal to 0. Therefore, we just solve this equation: $\qquad 2x - 6 = 0$

Adding 6 to both sides of the equation: $\qquad \dfrac{6 \quad 6}{}$

Results in: $\qquad 2x + 0 = 6$

Or just: $\qquad 2x = 6$

Dividing both sides by 2: $\qquad \dfrac{2x}{2} = \dfrac{6}{2}$

Results in: $\qquad x = 3$

This shows us that when $x = 3$, the denominator of $\dfrac{2x^2 + 3x - 5}{2x - 6}$ would

be 0 and the fraction would be undefined! Therefore in the fraction: $\qquad \dfrac{2x^2 + 3x - 5}{2x - 6}$

We would exclude 3 from possible values for x by indicating: $\qquad [x \neq 3].$

Example 4. **Multiply:** $\dfrac{x^2 - x - 20}{5x + 5} \cdot \dfrac{15x^2 - 15}{2x + 8}$

What values must be excluded from x for these fractions?
[When $5x + 5 = 0$; hence $x = -1$ or $2x + 8 = 0$; hence $x = -4$]

We use parentheses to indicate multiplication,

rewriting the expression: $\dfrac{\left(x^2 - x - 20\right)\left(15x^2 - 15\right)}{\left(5x + 5\right)\left(2x + 8\right)}$

Since there are no **identical** binomial or trinomial factors in their present form that can be removed as identity elements [canceled out], we will **factor as much as possible** to see what develops. First, we will

add the "missing" coefficient for the $-x$ term: $\dfrac{\left(x^2 - 1x - 20\right)\left(15x^2 - 15\right)}{\left(5x + 5\right)\left(2x + 8\right)}$

Each of the factors in both numerator and

denominator can be factored further: $\dfrac{\left(x - 5\right)\left(x + 4\right)\left(15\right)\left(x^2 - 1\right)}{5\left(x + 1\right) \cdot 2\left(x + 4\right)}$

We have not yet **factored completely** because the $\left(x^2 - 1\right)$ term is "the difference of perfect squares" and can

be factored into **conjugates**: $\dfrac{\left(x - 5\right)\left(x + 4\right)\left(15\right)\left(x + 1\right)\left(x - 1\right)}{5\left(x + 1\right)\left(2\right)\left(x + 4\right)}$

There are many **identical binomial factors** [identity elements] in the numerator and denominator and, therefore, can be removed. They can be removed because there is only the multiplication operation **between** every factor! To see them more clearly, we rearrange [commutative property] the factors to see

which ones cancel: $\dfrac{\overset{3}{\cancel{15}} \cdot \overset{1}{\cancel{(x+1)}} \cdot \overset{1}{\cancel{(x+4)}}(x-5)(x-1)}{\underset{1}{\cancel{5}} \cdot 2 \underset{1}{\cancel{(x+1)}} \cdot \underset{1}{\cancel{(x+4)}}}$

Removing the identity elements results in: $\dfrac{3(x - 5)(x - 1)}{2}$

[solution]

We must always keep in mind that $x=-1$ and $x=-4$ are excluded from possible values for x .

Division of Rational Expressions

First let's review how we divide regular fractions.

Example 5. [without variables] **Divide:** $\dfrac{6}{10} \div \dfrac{4}{3}$

To change **division** to **multiplication**, we take the **reciprocal**

of the **divisor**: $\dfrac{6}{10} \cdot \dfrac{3}{4}$

We then combine the fractions without actually performing

the multiplication: $\dfrac{6 \cdot 3}{10 \cdot 4}$

Canceling the common factor, 2, in both the numerator and denominator: $\dfrac{\cancel{6}^{3} \cdot 3}{\cancel{10}_{5} \cdot 4}$

Results in: $\dfrac{9}{20}$

[solution]

We use the same procedure for **division of rational expressions**.

Example 6. **Divide:** $\dfrac{2x^2}{4} \div \dfrac{x^4}{16}$

To change **division** to **multiplication**, we take the **reciprocal**

of the **divisor**: $\dfrac{2x^2}{4} \cdot \dfrac{16}{x^4}$

We combine the fractions: $\dfrac{2x^2 \cdot 16}{4 \cdot x^4}$

[Note: $2x^2$ can be viewed as $2 \cdot x^2$ since the coefficient, 2, is multiplying x^2]

And rearrange the factors: $\dfrac{16 \cdot 2 \cdot x^2}{4 \cdot x^4}$

We can rewrite x^4 in the denominator as its equivalent, $x^2 \cdot x^2$.

Canceling common factors:
$$\frac{\overset{4}{\cancel{16}} \cdot 2 \cdot \overset{\cdot 1}{\cancel{x^2}}}{\underset{1}{\cancel{4}} \cdot x^2 \cdot \underset{\cdot 1}{\cancel{x^2}}}$$

Resulting in:
$$\frac{8}{1x^2}$$

Or just:
$$\frac{8}{x^2}$$
[solution]

Example 7.   **Divide:**
$$\frac{10x+4}{x^2-4} \div \frac{5x^3+2x^2}{x+2}$$

To change **division** to **multiplication**, we take the **reciprocal** of the **divisor**:
$$\frac{10x+4}{x^2-4} \cdot \frac{x+2}{5x^3+2x^2}$$

We parenthesize the factors and combine the fractions:
$$\frac{(10x+4)(x+2)}{(x^2-4)(5x^3+2x^2)}$$

We now factor completely both the numerator and denominator. We start by further factoring $(10x+4)$ in the numerator:
$$\frac{2(5x+2)(x+2)}{(x^2-4)(5x^3+2x^2)}$$

We now factor (x^2-4) **[the difference of perfect squares]** in the denominator:
$$\frac{2(5x+2)(x+2)}{(x+2)(x-2)(5x^3+2x^2)}$$

We further factor $(5x^3+2x^2)$ in the denominator:
$$\frac{2(5x+2)(x+2)}{(x+2)(x-2)(x^2)(5x+2)}$$

We now cancel common factors:
$$\frac{2\cancel{(5x+2)}\cancel{(x+2)}}{\cancel{(x+2)}(x-2)(x^2)\cancel{(5x+2)}}$$

Resulting in:
$$\frac{2}{x^2(x-2)}$$
[solution]

Concept Homework

1. True or False:

a. In the algebraic fraction, $\dfrac{x^2 + 2x - 5}{2x - 5}$, we can cancel the identity element, $\dfrac{2x - 5}{2x - 5}$. If false, explain why this cannot be done.

b. In the algebraic fraction, $\dfrac{x^2(2x - 5)}{2x - 5}$, we can remove the identity element, $\dfrac{2x - 5}{2x - 5}$. If true, Explain why this can be done.

c. Is the following statement is **true**? $\dfrac{(x+5) - 6}{(x+5)} \Rightarrow \dfrac{-6}{1} = -6$. If not, what is the correct simplification?

d. Is the following statement is **true**? $\dfrac{(x+5)(-6)}{(x+5)} \Rightarrow \dfrac{-6}{1} = -6$. How does this example differ from example c.? What value for x is excuded from this simplification?

e. $\dfrac{(2x - 4)}{(2x - 4)}$ is a polynomial form of the identity element. If true, explain why.

Solutions: a. False; b. True; c. False; d. True; $x \neq -5$ e. True;

2. In order for the statement $\dfrac{x^2 + 2x - 5}{2x - 5} = x^2$ to be true, it would have to be true for **every value** assigned to x [except when $2x - 5 = 0$]. Assign a value to x of **your choice** [except $x = \frac{5}{2}$] to show that this canceling procedure is incorrect.

3. What value for x must be excluded from the fraction $\dfrac{x^2 + 2x - 5}{2x - 5}$ so that the fraction will never be undefined?

Solution: $x \neq \dfrac{5}{2}$

Exercises

Multiply/Divide and simplify.

1. $\dfrac{36}{65} \cdot \dfrac{20}{27}$ **2.** $\dfrac{9x^2}{4y^3} \cdot \dfrac{2y^2}{x}$ **3.** $\left(\dfrac{2x}{-y}\right)^3 \cdot \left(\dfrac{y}{3x}\right)^2$ **4.** $\dfrac{14x^2 - 21x}{24x - 16} \cdot \dfrac{12x - 8}{42x - 63}$

[Everything in each parenthesis
gets raised to the power indicated.]

5. $\dfrac{x^2 + 3x + 2}{x^4 y} \cdot \dfrac{x^3 y^2}{x^2 + 4x + 3}$ **6.** $\dfrac{4x^2 + 8x + 3}{6x^2 - x - 2} \cdot \dfrac{6x^2 - 7x + 2}{6x^2 + 7x - 3}$

7. $\dfrac{x^2 - 10x + 24}{30 + x - x^2} \cdot \dfrac{x^2 - 2x - 48}{x^2 - 12x + 32}$

$[\, 30 + x - x^2 \Rightarrow -\left(-30 - x + x^2\right) \Rightarrow (-1)\left(x^2 - x - 30\right) \,]$

8. $\dfrac{15}{26} \div \dfrac{45}{39}$ **9.** $\dfrac{4a^2 b^4}{9x^4 y^2} \div \dfrac{8a^4 b^9}{27x^3 y^6}$ **10.** $\dfrac{4x^3}{3x^2 - 3xy} \div \dfrac{x^2}{x^2 - y^2}$

11. $\dfrac{x^2 + 2x - 8}{x^2 - 3x - 4} \div \dfrac{x^2 - 4x + 4}{x^2 - 6x + 8}$ **12.** $\dfrac{10 + 9x - x^2}{x^2 - 7x - 8} \div \dfrac{x^2 - 8x - 20}{x^2 - 10x + 16}$

Solutions: **1.** $\dfrac{16}{39}$; **2.** $\dfrac{9x}{2y}$; **3.** $-\dfrac{8x}{9y}$; **4.** $\dfrac{x}{6}$; **5.** $\dfrac{y(x+2)}{x(x+3)}$; **6.** $\dfrac{2x-1}{3x-1}$; **7.** $-\dfrac{x+6}{x+5}$; **8.** $\dfrac{1}{2}$; **9.** $\dfrac{3y^4}{2a^2 b^5 x}$; **10.** $\dfrac{4(x+y)}{3}$;

11. $\dfrac{x+4}{x+1}$; **12.** $-\dfrac{x-2}{x+2}$.

Topic 23 - Addition and Subtraction with Rational Expressions (Algebraic Fractions) with Like Denominators

Addition of Rational Expressions with Like Denominators

When we add fractions, the denominators must be alike. Why is this? Consider the fraction $\frac{1}{5}$. We can think of this fraction as **"one of five equal parts"**. The **denominator** tells us how the whole is divided in a particular fraction. The **numerator** tells us how many of those equal parts are contained in the fraction. If we wanted to add two fractions, they both would have to be divided in the same way, hence the denominators would have to be the same. That is why to add (or subtract) fractions, the denominators must be alike. When adding fractions with like denominators, we **only add their numerators** and keep the denominator the same, since the denominator is only defining how the whole is divided.

Suppose we wanted to add $\frac{1}{5}$ and $\frac{2}{5}$. This would be like adding 1 of 5 equal parts to 2 of 5 equal parts. Logically the result would be 3 of 5 equal parts or $\frac{3}{5}$. Notice that the numerators are added and the denominators remain the same. $\frac{1}{5} + \frac{2}{5} = \frac{3}{5}$.

To generalize this concept for any fractions with like denominators, we would use variables:

$$\text{For addition: } \frac{a}{b} + \frac{c}{b} = \frac{a+c}{b} \text{ . For subtraction: } \frac{a}{b} - \frac{c}{b} = \frac{a-c}{b}$$

This process is the same with rational expressions (algebraic fractions) as it is for rational numbers without variables.

Example 1.

Add: $\frac{6x}{2n} + \frac{3x}{2n}$

These fractions are divided into $2n$ equal parts [provided that $n \neq 0$, **which would make the fractions undefined.**] Even though we don't know the actual value of $2n$, we know that it is the same for both fractions. Therefore, we can add the fractions together by adding their numerators.

Keeping the same denominator and adding the numerators: $\frac{6x + 3x}{2n}$

Combining like terms in the numerator results in: $\frac{9x}{2n}$

[Solution]

We do not know what the actual value of this fraction is, but we know that we performed the addition correctly since we have **followed the rules for fractional addition.**

Example 2. **Add and simplify:** $\dfrac{3y}{2z}+\dfrac{11y}{2z}$

Since we have common denominators, we add the numerators

and keep the denominator: $\dfrac{3y+11y}{2z}$

Combining like terms in the numerator results in: $\dfrac{14y}{2z}$

We can reduce this fraction by canceling out the common factor of 2: $\dfrac{\cancel{2}\cdot 7y}{\cancel{2}\cdot 1z}$

Resulting in: $\dfrac{7y}{z}$

[solution]

Example 3. **Add and simplify:** $\dfrac{x+8}{2x}+\dfrac{x+1}{2x}$

Since we have common denominators, we add the numerators

and keep the denominators: $\dfrac{(1x+8)+(1x+1)}{2x}$

We've added the implied "1" coefficients of the x terms in the numerators. We no longer need the parentheses in the

numerator since the binomials are being **added**: $\dfrac{1x+8+1x+1}{2x}$

Combining like terms in the numerator results in: $\dfrac{2x+9}{2x}$

We resist the temptation of canceling the $2x$ terms in the numerator and denominator since $2x$ and 9 in the numerator are **not factors** [multiplying each other]**.** They are connected by the **addition operation**. Therefore, the cancellation cannot be done

and the fraction is as simplified as possible: $\dfrac{2x+9}{2x}$

[solution]

It is difficult for some algebra students to understand why this cancelation discussed above is not possible. It might help your understanding to give a particular value to the x variable $[x \neq 0]$:

Suppose $x = 1$ in the fraction $\dfrac{2x+9}{2x}$. The fraction would then look like: $\dfrac{2(1)+9}{2(1)}$

Simplified, the fraction would be: $\dfrac{\frac{11}{2}}{1}$

Now let's look at the fraction if we performed the cancellation: $\dfrac{2\!\!\!/(1)+9}{2\!\!\!/(1)} \Rightarrow \dfrac{1+9}{1}$

Simplified, the value would then be: $\dfrac{\frac{10}{1}}{1}$

Or just: 10

Equivalent fractions **cannot have two different values** when $x = 1$. The correct evaluation is $\dfrac{11}{2}$ and not 10 because we cannot **cancel thru addition!**

Subtraction of Rational Expressions with Like denominators

We know how to change *any* subtraction problem to addition. We **change** the subtraction **operation** to addition by **changing the sign of the number (or expression) following the addition operation. We follow this same concept when dealing with rational expressions (algebraic fractions).**

Example 4. **Subtract and simplify:** $\dfrac{4x}{x-6} - \dfrac{24}{x-6}$

Now, we **change the subtraction operation to addition:** $\dfrac{4x}{x-6} + \dfrac{-24}{x-6}$

[**Note:** $\dfrac{-24}{x-6}$ has the opposite value of $\dfrac{24}{x-6}$ since the numerator's sign has been reversed while the denominator's sign has remained the same.]

Adding the numerators and keeping the denominator results in: $\dfrac{4x-24}{x-6}$

In order to simplify this fraction we will factor the numerator and see what develops. Factoring out 4 in the numerator results in:

$$\frac{4(x-6)}{x-6}$$

We now see that there is a **common binomial factor**, $(x-6)$, that can

cancel out:

$$\frac{4\cancel{(x-6)}}{\cancel{(x-6)}}$$

[$x-6 \;=\; (x-6)$ **when it stands alone in the denominator.**] The cancellation

Results in:

$$\frac{4}{1} \text{ or just } 4$$

[Solution]

[**What value of** x **is excluded from this subtraction operation? By now you know that** $x \neq 6$ **since that value for x would make the denominator zero and the fraction undefined.**]

You might be wondering why we can cancel out the $(x-6)$ in the numerator and

denominator in this example and could not cancel $2x$ in $\dfrac{2x+9}{2x}$ the previous example.

This is because $(x-6)$ is a **factor** in the numerator, $4(x-6)$ [**It is connected to 4 through**

multiplication, not addition.] There is an **identical** binomial factor in the denominator $\dfrac{4(x-6)}{(x-6)}$,

so they can be canceled.

It might be helpful to give a specific example to show that $\dfrac{4(x-6)}{x-6} \;=\; 4$. If we assign 1 as

the value of x , this equation would be $\dfrac{4(1-6)}{1-6} \;=\; 4$. Is this a true statement? Simplifying,

we get $\dfrac{4(-5)}{-5} = 4$. Continuing with the simplification results in $\dfrac{-20}{-5} = 4$ which is true! You

will also find that it is true for any value of x except when $x=6$. Why is that value

excluded? [**It would make the denominator equivalent to 0.**]

Here is an example that requires **trinomial factoring**:

Example 5. **Subtract and simplify:** $\dfrac{5-3a}{a^2-2a+1} - \dfrac{a+1}{a^2-2a+1}$

We change the subtraction operation to addition: $\dfrac{5-3a}{a^2-2a+1} + \dfrac{-(a+1)}{a^2-2a+1}$

Notice that we parenthesized $a+1$ in the second fraction. This is because the **entire** numerator must have its sign reversed, not just the a term! Adding the numerators and

keeping the denominator:

$$\frac{5-3a-(a+1)}{a^2-2a+1}$$

Simplifying the numerator results in:

$$\frac{5-3a-1a-1}{a^2-2a+1}$$

[Notice that *all* the signed were reversed in $(a+1)$]

Combining like terms in the numerator results in:

$$\frac{4-4a}{a^2-2a+1}$$

Since we cannot cancel through addition or subtraction, we will factor the numerator and denominator to see

what develops:

$$\frac{4(1-a)}{(a-1)(a-1)}$$

There is no identical factors to be canceled however we will now utilize a very useful algebraic tool! Is there a relationship between $\frac{(1-a)}{(a-1)}$? There answer is **yes!** Let's factor out -1

from the numerator, $(1-a)$: $-1\left(\frac{1}{-1}+\frac{-1a}{-1}\right)$ results in

$-1(-1+a)$. Notice that factoring out -1 results in **reversing signs of each term in the parenthesis.** Now let's use the commutative property of addition to reorder the terms in the parenthesis results in $-1(a-1)$. So $\boxed{(1-a)=-1(a-1)}$

In this situation, this becomes very useful, because it creates **identical binomial factors** in the numerator and

denominator of the fraction that we are simplifying:

$$\frac{4(1-a)}{(a-1)(a-1)}$$

Becomes:

$$\frac{4(-1)(a-1)}{(a-1)(a-1)}$$

The entire fraction is now made up of factors, so we can

cancel the **identical binomial factors:**

$$\frac{4(-1)\,\cancel{(a-1)}}{(a-1)\,\cancel{(a-1)}}$$

Resulting in:

$$\frac{4(-1)}{(a-1)}$$

Simplifying the numerator:

$$\frac{-4}{(a-1)}$$

Or more conventionally:

$$-\frac{4}{(a-1)}$$

[solution]

This shows us that $\dfrac{5-3a}{a^2-2a+1} - \dfrac{a+1}{a^2-2a+1}$ **is equivalent to** $-\dfrac{4}{(a-1)}$ for all values

assigned to a except such values that will make the either of the denominators equal to 0.

You can test the validity of the equation, $\dfrac{5-3a}{a^2-2a+1} - \dfrac{a+1}{a^2-2a+1} = -\dfrac{4}{(a-1)}$ by

randomly assigning a value to a to show that it makes **a true statement.** You will find that the equation will be true for all values of a except those values that make either denominator equivalent to 0 [$x \neq 1$].

Concept Homework

1. Give an example of an identity element that contains binomials.

2. Why can you cancel binomials in the fraction $\dfrac{c(a+b)}{d(a+b)}$, but not in the fraction $\dfrac{c+(a+b)}{d+(a+b)}$?

3. Assign the values of your choice for a, b, c and d to show that $\dfrac{c(a+b)}{d(a+b)} = \dfrac{c}{d}$.

   [Keep in mind that you cannot use values that makes the denominator equivalent to zero.]

4. Assign the values of your choice for a, b, c and d to show that $\dfrac{c+(a+b)}{d+(a+b)} \neq \dfrac{c}{d}$.

5. Show how $\dfrac{(x-y)}{(y-x)}$ can be simplified.

212

Exercises

Perform the indicated operation and simplify you answer as much as possible:

1. $\dfrac{6}{3x} + \dfrac{5}{3x}$

2. $\dfrac{3}{7} + \dfrac{2}{7} - \dfrac{y}{7}$

3. $\dfrac{2}{5x} - \dfrac{9}{5x} - \dfrac{3}{5x}$

4. $\dfrac{x-2y}{3x} + \dfrac{x+3y}{3x}$

5. $\dfrac{x-2}{a+b} - \dfrac{2x+1}{a+b}$

6. $\dfrac{2a-b}{b} - \dfrac{a-2b}{b}$

7. $\dfrac{x}{2x+4} - \dfrac{2-x}{2x+4}$

8. $\dfrac{x+y}{2(x-y)} + \dfrac{2x-2y}{2(x-y)} + \dfrac{x-3y}{2(x-y)}$

9. $\dfrac{5}{2x-5} - \dfrac{2x}{2x-5}$

10. $\dfrac{4}{3x^2-5x-2} - \dfrac{x^2}{3x^2-5x-2}$

   Hint: add the numerators. Then factor
   both the numerator and denominator.

11. $\dfrac{x^2-3}{x^2-8x+12} + \dfrac{2x-1}{x^2-8x+12} - \dfrac{x+2}{x^2-8x+12}$

Solutions. 1. $\dfrac{11}{3x}$; 2. $\dfrac{5-y}{7}$; 3. $-\dfrac{2}{x}$; 4. $\dfrac{2x+y}{3x}$; 5. $\dfrac{-x-3}{a+b}$ or $-\dfrac{(x+3)}{a+b}$; 6. $\dfrac{a+b}{b}$; 7. $\dfrac{x-1}{x+2}$; 8. 2 ; 9. -1 ;

10. $\dfrac{-2-x}{3x+1}$ or $-\dfrac{x+2}{3x+1}$; 11. $\dfrac{x+3}{x-6}$

Topic 24 - Addition and Subtraction with Rational Expressions (Algebraic Fractions) with Unlike Denominators

As discussed in the previous topic, addition and subtraction of **any** fractions require that the denominators be the same. Let's look at the situation where denominators are not the same. We will start with arithmetic fractions as the rules will be the same for algebraic fractions. Consider the fraction $\frac{1}{2}$ [one of two equal parts]. If we wanted to add $\frac{1}{2}$ to $\frac{2}{3}$, [two of three equal parts], we couldn't do this because the division of "the whole" is not the same with these fractions. Somehow, we would have to change how "the wholes" are broken up **without changing *the value* that each fraction represents**. We do this by finding "**a common denominator**".

This process is the same, whether the fractions contain variables or not!

Example 1 [without variables]. **Add:** $\frac{1}{2}+\frac{2}{3}$

How do we change 2 equal parts of the whole and 3 equal parts of the whole so that they have the same denominators **without changing the value of the fractions themselves? We find *convenient* identity elements to multiply each fraction!**

[Remember: Multiplying by an identity element doesn't change a number's value, because all identity elements for multiplication have a value of 1. Multiplying by 1 doesn't change the value of a number.]

Which **configurations** of the identity elements do we choose?
The nice thing is that **we can *choose any configuration we want*:** $\frac{3}{3}\cdot\frac{1}{2}+\frac{2}{2}\cdot\frac{2}{3}$

The identity elements, $\frac{3}{3}$ and $\frac{2}{2}$ [which are each equivalent to "1"] are

convenient because they make the denominators in both cases, the same $[6]$: $\frac{3}{6}+\frac{4}{6}$

The fractions $\frac{3}{6}$ and $\frac{4}{6}$ have the **same value** as the original fractions

$\frac{1}{2}$ and $\frac{2}{3}$ and they now have the same division of the whole

[common denominators]. Because of this we can now add them: $\frac{7}{6}$
[solution]

Note: In this case, the common denominator turns out to be the original denominators multiplied together! Multiplying the original denominators

will *always* result in a common denominator, although not always the lowest possible common denominator or what we call the "least common denominator" [LCD]. The same process occurs when we adding fractions that have variables in the denominators:

Example 2.

Add: $\dfrac{1}{6}+\dfrac{2}{n}$

The whole of the first fraction is broken into 6 equal parts and the whole of the second fraction is broken into "n" equal parts. Now, as with regular fractions, we must find **convenient** identity elements to multiply each fraction to create a common denominator!
[Remember: Multiplying by identity elements doesn't change a fraction's value.]

Which identity elements do we choose? The nice thing is that we can ***choose any ones we want***: $\dfrac{n}{n}\cdot\dfrac{1}{6}+\dfrac{6}{6}\cdot\dfrac{2}{n}$

The identity elements, $\dfrac{n}{n}$ $\left[\text{yes, }\tfrac{n}{n}=1\right]$ and $\dfrac{6}{6}$ are **convenient** because they make the denominators in both cases, the same $[6n]$: $\dfrac{n}{n\cdot6}+\dfrac{12}{6\cdot n}$

Is $n\cdot6$ and $6\cdot n$ the same quantity? Yes! They both have the same value: $6n$. This is the **commutative property of multiplication** in action: $\dfrac{n}{6n}+\dfrac{12}{6n}$

Therefore, we can add these fractions because they are both broken into $6n$ parts: $\dfrac{n+12}{6n}$
[solution]

[We can't combine n and 12 in the numerator because they are not "like terms". They must remain as they are.]

Note: The common denominator, in this case, are original denominators multiplied together.

Example 3.

Add: $\dfrac{x}{y}+\dfrac{y}{x}$

Our first impulsemight be to cancel the x's and y's in the numerators and denominators, thereby arriving at an **incorrect** solution. **We can only cancel factors** [terms being multiplied] in an expression. Since this problem involves the addition **operation**, we must instead find a common denominator so that we can add the fractions. We can always arrive at a common denominator by **multiplying** the denominators.

The first fraction is multiplied by $\dfrac{x}{x}$ [**an identity element**] and the

second is multiplied by $\dfrac{y}{y}$ [**also an identity element**] creating the

common denominator, xy **or** yx **which are equivalent**: $\quad \dfrac{x}{x}\cdot\dfrac{x}{y} + \dfrac{y}{y}\cdot\dfrac{y}{x}$

Therefore, we can rewrite the second fraction: $\quad \dfrac{xx}{xy} + \dfrac{yy}{xy}$

Since the denominators of each fraction are the same, we can add

the numerators: $\quad \dfrac{x^1x^1 + y^1y^1}{xy}$

[We show the implied exponents of x and y in the numerator.]

Using our rule for multiplying exponents with the same base: $\quad \dfrac{x^2 + y^2}{xy}$

[solution]

Example 4. **Add:** $\quad \dfrac{7x^2}{(x-4)^2} + \dfrac{5x}{3x-12}$

This is a case where multiplying the denominators would become quite messy and will not give us the **least** common denominator. How can we find the LCD with algebraic fractions? We can find the LCD

by **factoring each denominator**: $\quad \dfrac{7x^2}{(x-4)(x-4)} + \dfrac{5x}{3(x-4)}$

Upon close inspection, we see that the first denominator is missing the factor "3" and the second fraction requires an additional $(x-4)$ factor to make both denominators equivalent as well as LCD's. So, we multiply each fraction by the necessary identity

elements to accomplish this: $\quad \dfrac{3}{3}\cdot\dfrac{7x^2}{(x-4)(x-4)} + \dfrac{5x}{3(x-4)}\cdot\dfrac{(x-4)}{(x-4)}$

[$\dfrac{x-4}{x-4}$ is an identity element! In fact any fraction with

identical numerators and denominators is an identity element.]

Now we have common denominators: $3(x-4)(x-4)$ in each fraction. That allows us to add the numerators

while keeping the common denominator:

$$\frac{3(7x^2)+5x(x-4)}{3(x-4)(x-4)}$$

[Note: We are not able to do any canceling at this point because of the addition operation in the numerator.]
Carrying out the multiplication and distribution

operations in the numerator:

$$\frac{21x^2+5x^2-20x}{3(x-4)(x-4)}$$

Adding like terms in the numerator:

$$\frac{26x^2-20x}{3(x-4)(x-4)}$$

Finally, we factor the numerator to see if there are any common factors that can be canceled. In this

case there are not any:

$$\frac{2x(13x-10)}{3(x-4)(x-4)}$$

[solution]

Subtraction of algebraic fractions with unlike denominators

We know how to change **any** subtraction problem to addition. We just **change** the subtraction **operation** to addition and change the **sign** of the number [or expression] following it. **We follow this same concept when dealing with algebraic fractions.**

Example 5. **Subtract:** $\dfrac{3}{y-2} - \dfrac{5}{y}$

First, we will change the subtraction operation to addition: $\dfrac{3}{y-2} + \dfrac{-5}{y}$

[Note: $\dfrac{-5}{y}$, has a the opposite value of $\dfrac{5}{y}$]

Although the denominators appear to be unusual [one is a binomial and the other is a monomial], we have learned from previous examples that we can always find a common denominator by multiplying the original denominators. Since neither denominator can be factored this process will yield the **least common denominator** (LCD).Therefore the common denominator that we seek is $y(y-2)$. We can multiply the first fraction by the identity element $\dfrac{y}{y}$ and the second fraction by the identity element $\dfrac{(y-2)}{(y-2)}$ **without changing their values** since they are **both identity elements**. This results in :

$$\frac{y}{y} \cdot \frac{3}{(y-2)} + \frac{-5}{y} \cdot \frac{(y-2)}{(y-2)}$$

Now the fractions have the **same denominators** $[y(y-2)]$:

$$\frac{3y}{y(y-2)} + \frac{-5(y-2)}{y(y-2)}$$

So we can add the numerators of these fractions, keeping

the common denominator:

$$\frac{3y+(-5)(y-2)}{y(y-2)}$$

To **simplify the fraction**, we distribute in the numerator:

$$\frac{3y-5y+10}{y(y-2)}$$

Combining like terms in the numerator:

$$\frac{-2y+10}{y(y-2)}$$

Factoring out -2 in the numerator results in:

$$\frac{-2(y-5)}{y(y-2)}$$

[solution]

There are no common factors that can be canceled.

This solution could be expressed as:

$$-\frac{2(y-5)}{y(y-2)}$$

[solution]

When a denominator can be factored, this should be done to determine the least common denominator as we will do below in the next example:

Example 6.   **Subtract:**
$$\frac{1}{x-3} - \frac{6}{x^2-9}$$

We start by changing the subtraction of the fractions to addition:
$$\frac{1}{x-3} + \frac{-6}{x^2-9}$$

We notice that the second faction's denominator can be factored:
$$\frac{1}{(x-3)} + \frac{-6}{(x+3)(x-3)}$$

Now we see that the **least** common denominator is $(x+3)(x-3)$.

The first fraction only requires a factor of $(x+3)$ to be equivalent to the second fraction. We can accomplish this by multiplying

the first fraction by $\frac{(x+3)}{(x+3)}$, the necessary identity element:
$$\frac{(x+3)}{(x+3)} \cdot \frac{1}{(x-3)} + \frac{-6}{(x+3)(x-3)}$$

Now, keeping the common denominator we add the numerators:
$$\frac{(x+3)+(-6)}{(x+3)(x-3)}$$

[Notice that $(x+3)\cdot 1 = (x+3)$, since multiplying by 1 doesn't change anything]

There is no longer a need for the parenthesis indicating the binomial, $(x+3)$, in the numerator since all the operations in the numerator are addition:

$$\frac{x+3+(-6)}{(x+3)(x-3)}$$

Combining the like terms [constants] in the numerator results in:

$$\frac{x-3}{(x+3)(x-3)}$$

Removing the identity element, $\frac{x-3}{(x-3)} = \frac{1}{1}$ gives us the simplified solution:

$$\frac{1}{(x+3)}$$

[solution]

Notice that the numerator did not just "disappear". If there is nothing for the identity element, "1" to multiply, we must keep it in the numerator.

Concept Homework

As an assessment of your understanding of the concepts set forth in this section, answer the following questions. If necessary, review the material to help you arrive at the correct conclusions. With "true or false" statements, if the statement is true, make up an example that supports the statement without using examples given in the material. If a statement is false, correct the wording to make it true. Then make up an example that supports the statement without using examples given in the material.

1. **True or False**

   a. **Identity elements can contain variables, as long as they are fractions with equivalent numerators and denominators.**

   b. **We can always arrive at a common denominator with algebraic fractions by multiplying the original denominators together.**

   c. **With the fractions $\dfrac{a}{b} \cdot \dfrac{b}{a}$, we can "cancel" the a's and b's $_{[a, b \neq 0]}$. This results in the identity element for multiplication, 1 .**

   d. **With the fractions $\dfrac{a}{b} + \dfrac{b}{a}$, we can "cancel" the a's and b's . This results in the equivalent fractions $\dfrac{1}{1} + \dfrac{1}{1}$ or $1 + 1$. Therefore, $\dfrac{a}{b} + \dfrac{b}{a} = 2$.**

   Solutions : 1a. True; 1b. True; 1c. True; 1d. False

2. **Show that if $a = 1$ and $b = 2$, $\dfrac{a}{b} + \dfrac{b}{a} \neq 2$** [Show that this would not be a true statement].

3. **Assign the values of your choice for a and b $_{[a, b \neq 0]}$ to show that $\dfrac{a}{b} + \dfrac{b}{a}$ will not always equal 2 .**

4. **Assign the values of your choice $_{[a, b \neq 0]}$ for a and b to show that $\dfrac{a}{b} \cdot \dfrac{b}{a}$ will always result in 1 .**

Exercises

Add or Subtract. Express the solution in simplified form.

1. $\dfrac{7}{2}+\dfrac{3}{8}$ **2.** $\dfrac{7}{2}+\dfrac{3}{x}$ **3.** $\dfrac{7}{2x}+\dfrac{5}{x^2}-\dfrac{3}{4x}$ **4.** $\dfrac{x+6}{9x^2}-\dfrac{3x-4}{6x^2}$

5. $\dfrac{2x}{x^2-9}-\dfrac{2}{x+3}-\dfrac{1}{x-3}$ **6.** $\dfrac{x+1}{x^2+x-12}-\dfrac{2}{x^2+5x-24}$

[**Hint:** $-x+3 \Rightarrow (3-x) \Rightarrow -(x-3)$]

7. $\dfrac{1}{x-4}-\dfrac{1}{3-x}+\dfrac{1}{x^2-7x+12}$

[**Hint:** $(b-a)=-(a-b)$]

Solutions: **1.** $\dfrac{31}{8}$; **2.** $\dfrac{7x+6}{2x}$; **3.** $\dfrac{11x+20}{4x^2}$; **4.** $\dfrac{-7x+24}{18x^2}$; **5.** $\dfrac{-1}{x+3}$ or $-\dfrac{1}{x+3}$; **6.** $\dfrac{x(x+7)}{(x+4)(x-3)(x+8)}$; **7.** $\dfrac{2}{(x-4)}$

Topic 25 - Solving Equations with Rational Expressions (Algebraic Fractions)

Adding algebraic fractions with unlike denominators is closely related to solving equations with algebraic fractions. However, instead of creating a least common denominator for adding the fractions, **we will multiply every term in the equation by what would be the least common denominator.** This process will remove all the denominators from the equation through canceling, making its solution simpler. We will start by reviewing the addition process and then showing how it differs from solving equations that contain fractions.

Example 1.   **Add**: $\dfrac{x}{2} + \dfrac{x}{3}$

We need to find a common denominator for the fractions. Since both denominators are **prime numbers,** the least common denominator is found by multiplying them which results in an LCD of 6. To create this denominator without changing the value of each fraction, we choose an identity element that will make both denominators 6 .

$[\dfrac{3}{3}$ for the first fraction and $\dfrac{2}{2}$ for the second]: $\dfrac{3 \cdot x}{3 \cdot 2} + \dfrac{2 \cdot x}{2 \cdot 3}$

Having created the common denominator of 6 , we now add the

numerators and keep the denominator, resulting in: $\dfrac{3x + 2x}{6}$

Adding like terms in the numerator, the fraction becomes: $\dfrac{5x}{6}$

[Solution]

Example 2.

Now suppose we wanted to **solve an equation** for x

containing **algebraic fractions:** $\dfrac{x}{2} + \dfrac{x}{3} = 10$

Since we are now dealing with **an equation**, the objective is to find a value(s) for x that will **make the equation true**. This process will be simplified if we eliminate the denominators in the equation **without changing the equivalence of the right and left sides of the equation.**

We see that the least common denominator of the fractions is 6 . **Rather than changing the dominators** to 6 , the most **efficient approach** is to **multiply every term in the equation by** 6 . This will **not change** the "balance" of the equation:

$$\frac{6}{1}\left(\frac{x}{2}+\frac{x}{3}\right)=6\cdot 10$$

We can use $\frac{6}{1}$ [which is equivalent to 6] when the terms we are multiplying are fractions, and its equivalent, 6 , when we are multiplying non-fractional terms. Performing the multiplication results in:

$$\frac{6x}{2}+\frac{6x}{3}=60$$

The advantage to this approach is that **all the denominators will "cancel out" of the equation.** Factoring 6 into $2\cdot 3$ **and** $3\cdot 2$:

$$\frac{\cancel{2}\cdot 3x}{\cancel{2}}+\frac{\cancel{3}\cdot 2x}{\cancel{3}}=60$$

Removing the identity elements, results in:

$$\frac{3x}{1}+\frac{2x}{1}=60$$

Since dividing by 1 , the **identity element** for division, doesn't change the value of the fraction, we can express the fractions as a whole numbers without the denominator:

$$3x+2x=60$$

Now we are dealing with a simple equation **without fractions!**

Combining like terms:

$$5x=60$$

dividing both sides of the equation by 5 :

$$\frac{5x}{5}=\frac{60}{5}$$

Results in:

$$x=12$$
[solution]

To check that our solution is correct, we substitute our solution, $x=12$, into the original equation, $\frac{x}{2}+\frac{x}{3}=10$, to see if it makes

The equation true:

$$\frac{12}{2}+\frac{12}{3}=10$$

Simplifying the fractions:

$$6+4=10$$

We see that $x=12$ **makes the equation true,** and our solution is correct.

Notice that we did not re-configure the fractions so that they had a least common denominator. Instead we **multiplied every term by the least common denominator** and were then able to **eliminate the denominators by "canceling".**

Sometimes equations with fractions are presented in a confusing way. Don't panic! Just **"un-confuse"** them!

Example 3. **Solve for** x **:** $\dfrac{1}{3}(x+2) - \dfrac{x}{4} = 2$

The first term, $\dfrac{1}{3}(x+2)$, is confusing. We can clarify it by expressing $(x+2)$ as a fraction by putting it over 1, which

doesn't change its value: $\dfrac{1}{3} \cdot \dfrac{(x+2)}{1} - \dfrac{x}{4} = 2$

After multiplying the first fractions, the equation becomes: $\dfrac{(x+2)}{3} - \dfrac{x}{4} = 2$

The problem is now in a more familiar form. Using the **previous strategy**, we **multiply** *every term* **in the equation** by the **least common denominator** of the fractions $[\frac{12}{1}$ or $12\,]$: $\dfrac{12}{1}\left(\dfrac{(x+2)}{3} - \dfrac{x}{4} \right) = 12 \cdot 2$

Distributing and multiplying results in: $\dfrac{12(x+2)}{3} - \dfrac{12 \cdot x}{4} = 24$

We can now factor and eliminate identity elements **["cancel"]:** $\dfrac{\cancel{3} \cdot 4(x+2)}{\cancel{3}} - \dfrac{\cancel{4} \cdot 3x}{\cancel{4}} = 24$

Resulting in: $\dfrac{4(x+2)}{1} - \dfrac{3x}{1} = 24$

We no longer need the denominators, since they are 1's : $4(x+2) - 3x = 24$

Distributing the "4" results in: $4x + 8 - 3x = 24$

Adding like terms: $1x + 8 = 24$

Adding -8 to both sides: $\underline{\quad -8 \quad -8\quad}$

Results in: $x \quad = 16$

[solution]

We can **check your solution** by substituting 16 for x

into the original equation: $\dfrac{1}{3}(x+2) - \dfrac{x}{4} = 2$

This results in: $\dfrac{1}{3}(16+2) - \dfrac{16}{4} = 2$

Performing the addition in the parenthesis first and putting the

Result over 1: $\quad \dfrac{1}{3}\left(\dfrac{18}{1}\right) - \dfrac{16}{4} = 2$

Next, we perform the multiplication: $\quad \dfrac{18}{3} - \dfrac{16}{4} = 2$

Simplifying the fractions: $\quad 6 - 4 = 2$

Since $6 - 4 = 2$ **is a true statement,** we know our answer is correct!

Sometimes it is **not efficient** to multiply algebraic denominators together to find a common denominator. It is **more efficient** to **factor the denominators** to find the **least common denominator:**

Example 4. $\qquad$ **Solve for** x: $\quad \dfrac{4x}{x^2 + x - 30} + \dfrac{2}{x - 5} = \dfrac{1}{x + 6}$

We certainly don't want to multiply all these denominators together to find a common denominator if we don't have to! Instead, we will **factor** the first denominator and put parentheses around the other denominators since they will be

treated as **factors**: $\quad \dfrac{4x}{(x-5)(x+6)} + \dfrac{2}{(x-5)} = \dfrac{1}{(x+6)}$

We notice that **all** the denominators are made up of the same factors: $(x-5)$ and $(x+6)$. Our **least** common denominator, therefore, is only $(x-5)(x+6)$ since every denominator will cancel out when their fractions are multiplied by $(x-5)(x+6)$. So we **multiply the entire equation** by $(x-5)(x+6)$ which it the **least** common denominator. This multiplication looks like:

$$\frac{(x-5)(x+6)}{1} \cdot \frac{4x}{(x-5)(x+6)} + \frac{(x-5)(x+6)}{1} \cdot \frac{2}{(x-5)} = \frac{(x-5)(x+6)}{1} \cdot \frac{1}{(x+6)}$$

We now "cancel" the identity elements:

$$\frac{\cancel{(x-5)}\,\cancel{(x+6)}}{1} \cdot \frac{4x}{\cancel{(x-5)}\,\cancel{(x+6)}} + \frac{\cancel{(x-5)}\,(x+6)}{1} \cdot \frac{2}{\cancel{(x-5)}} = \frac{(x-5)\,\cancel{(x+6)}}{1} \cdot \frac{1}{\cancel{(x+6)}}$$

Resulting in: $\dfrac{4x}{1}+\dfrac{(x+6)}{1}\cdot\dfrac{2}{1}=\dfrac{(x-5)}{1}$

Since all the denominators are 1's, they can "disappear": $4x+(x+6)\cdot 2=(x-5)$

Using the commutative property to re-arrange the factor, 2, in the second term: $4x+2\cdot(x+6)=(x-5)$

Distributing: $4x+2x+12=x-5$

Combining like terms and adding a "1" coefficient to the x on the right side: $6x+12=1x-5$

We now want to get all the x variables on one side of the equation. We add $-1x$ to both sides of the equation: $\quad\dfrac{-1x\qquad\quad -1x}{}$

Resulting in: $5x+12=\quad -5$

We isolate the $5x$ term by adding -12 to both sides of the equation: $\dfrac{-12\qquad -12}{}$

Resulting in: $5x\quad=\quad -17$

Dividing both sides of the equation by 5: $\dfrac{\cancel{5}x}{\cancel{5}}=\dfrac{-17}{5}$

Results in: $1x=\dfrac{-17}{5}$

Or more conventionally: $x=-\dfrac{17}{5}$

[solution]

Here is an **application** for equations with algebraic fractions:

Example 4.

If one third is deducted from four times the reciprocal of y, the result is 6. What is the value of y?

We must **decipher** the information given and **present it as an equation**.

1. y [equivalent to $\left[\dfrac{y}{1}\right]$] has a reciprocal of $\dfrac{1}{y}$.

2. Four times this is $\dfrac{4}{1}\cdot\dfrac{1}{y}$. This results in $\dfrac{4}{y}$. Therefore,

"**four times the reciprocal of y**" is $\dfrac{4}{y}$.

3. The problem states that $\dfrac{1}{3}$ is **deducted [subtracted] from** this amount and the **result is** 6. "The result is" can be interpreted mathematically as $"="$.

We can now write the equation:
$$\frac{4}{y} - \frac{1}{3} = 6$$

Neither denominator can be factored, so the LCD is found by **multiplying** the denominators. Therefore, LCD for these fractions is $3y$. Following our **previous strategy**, we will

Multiply **every term** in the equation by $\frac{3y}{1}$ or $3y$:
$$\frac{3y}{1}\left(\frac{4}{y} - \frac{1}{3}\right) = 3y(6)$$

Distributing results in:
$$\frac{3y}{1} \cdot \frac{4}{y} - \frac{3y}{1} \cdot \frac{1}{3} = 3 \cdot y \cdot 6$$

The multiplication results in:
$$\frac{12y}{1y} - \left(\frac{3y}{3}\right) = 18y$$

Now we can simplify these terms by canceling:
$$\frac{12 \cdot \cancel{y}}{1\cancel{y}} - \frac{\cancel{3} \cdot y}{\cancel{3}} = 18y$$

Resulting in:
$$\frac{12}{1} - \frac{y}{1} = 18y$$

We can now eliminate the "1" denominators:
$$12 - y = 18y$$

Now we can add the "1" coefficient for the y term on the Left side:
$$12 - 1y = 18y$$

To isolate the y terms on the right side of the equation, We add $1y$ to both sides of the equation:

Resulting in:
$$12 \quad \frac{1y \qquad 1y}{ = 19y}$$

Dividing both sides of the equation by 19:
$$\frac{12}{19} = \frac{\cancel{19}y}{\cancel{19}}$$

Results in:
$$\frac{12}{19} = y$$

Or, more conventionally:
$$y = \frac{12}{19}$$
[solution]

Concept Homework

As an assessment of your understanding of the concepts set forth in this section, answer the following questions. If necessary, review the material to help you arrive at the correct conclusions. With "true or false" statements, if the statement is true, make up an example that supports the statement without using examples given in the material. If a statement is false, correct the wording to make it true. Then make up an example that supports the statement without using examples given in the material.

True or False

1. **When solving equations with algebraic fractions the most efficient strategy is to create a common denominator for all of the fractions.**

2. **Multiplying each term in an equation that contains fractions by the LCD will allow us to eliminate the denominators of the fractions throughout the equation by "canceling".**

3. $\dfrac{1}{4}(3y^2 - 2)$ **is equivalent to** $4(3y^2 - 2)$**.**

4. **If we multiply every term in the equation** $\dfrac{y}{9} + \dfrac{2}{y} - \dfrac{3y-4}{3} = 10$ **by** $9y$**, we will be able to eliminate the denominators from the equation.**

5. **The reciprocal of** x **is** $\dfrac{x}{1}$**.**

6. **5 times the reciprocal of** y **is** $\dfrac{5}{y}$**.**

7. **With the following expression,** $\dfrac{x}{4} + \dfrac{9}{x}$**, we can cancel the** x's **.**

8. **With the following expression,** $\dfrac{x}{4} \cdot \dfrac{9}{x}$**, we can cancel the** x's **.**

Solutions: 1. False; 2. True; 3. False; 4. True; 5. False; 6. True; 7. False; 8. True.

228

Exercises

Solve each equation.

1. $\dfrac{5x}{2} = x + \dfrac{1}{4}$ 2. $\dfrac{5y}{6} - \dfrac{1}{6} = \dfrac{2y}{3} + \dfrac{5}{6}$ 3. $1 + \dfrac{3}{x} = \dfrac{12}{x}$ 4. $\dfrac{y+12}{9} = \dfrac{y-9}{2}$

5. $\dfrac{x-1}{10} + \dfrac{19}{15} = \dfrac{x}{3}$ 6. $\dfrac{2}{x+10} = \dfrac{1}{x+3}$ 7. $\dfrac{4}{y^2 - y - 20} = \dfrac{-5}{y+4}$.

Create an equation that mathematically represents each problem and then solve the equation.

8. If one-half of a certain number is added to three times the number, the result is $\dfrac{35}{2}$. Find the number.

9. If two-thirds is deducted from the three times the reciprocal of the number, the result is 20. Find the number.

10. The denominator of a number is six more than the numerator and the fraction is equivalent to $\dfrac{3}{4}$. Find the number.

Solutions: **1.** $x = \dfrac{1}{6}$; **2.** $y = 6$; **3.** $x = 9$; **4.** $y = 15$; **5.** $x = 5$; **6.** $x = 4$; **7.** $y = \dfrac{21}{5}$; **8.** $x = 5$; **9.** $x = \dfrac{9}{62}$; **10.** $x = 18$;

228

Topic 26 – Using Proportions to Solve Equations with Rational Expressions

A proportion is an equation made up of two ratios. To say that two ratios are proportional means that one ratio is quantitatively equivalent to the other. For example, $\frac{1}{3} = \frac{2}{6}$. This is a true statement since $\frac{2}{6}$ can be reduced to $\frac{1}{3}$ by factoring and canceling $\left[\frac{2}{6} \Rightarrow \frac{1 \cdot \cancel{2}}{3 \cdot \cancel{2}} = \frac{1}{3} \right]$.

We can have a rational expression in a proportion. For example, $\frac{5}{x} = \frac{2}{10}$. This now **becomes an equation** that can be solved for the variable, x.

Let's examine how we easily remove the fractions from **all** proportions.

We can do this by using variables to generalize **any** proportion: $\frac{a}{b} = \frac{c}{d}$

To remove the fractions, we can multiply both sides of the equation by the common denominator, bd [**which, of course, means** $b \cdot d$ **]** expressed as a

fraction: $\frac{bd}{1} \cdot \frac{a}{b} = \frac{bd}{1} \cdot \frac{c}{d}$

Combining the fractions through multiplication results in: $\frac{bda}{1b} = \frac{bdc}{1d}$

Using the commutative property of multiplication we can express the variables of each fraction in alphabetical order: $\frac{abd}{1b} = \frac{bcd}{1d}$

Removing the identity elements for multiplication for both fractions: $\frac{a\cancel{b}d}{1\cancel{b}} = \frac{bc\cancel{d}}{1\cancel{d}}$

Results in: $\frac{ad}{1} = \frac{bc}{1}$

We no longer need to express these terms as fractions, Since their denominators are "1": $ad = bc$

These algebraic manipulations show us that, **in all cases**:

$$\boxed{\frac{a}{b} = \frac{c}{d} \text{ is mathematically equivalent to } ad = bc}$$

Putting this into words:

The numerator of the left side fraction times the denominator of the right side fraction will always be equal to
the denominator of the left side fraction times the numerator of the right side fraction.

Mathematically speaking we say:

In proportions, ***cross products*** are equal.

Using the example, $\dfrac{1}{3} = \dfrac{2}{6}$ we can **restate this proportion** using **cross products**:

$$1 \cdot 6 = 3 \cdot 2$$

We will now use this **cross product concept** to solve proportional equations.

Example 1.   Solve for x: $\dfrac{3}{15} = \dfrac{5}{x}$

Since cross products of a proportion are equal: $3 \cdot x = 15 \cdot 5$

Multiplying: $3x = 75$

Dividing both sides of the equation by 3: $\dfrac{3x}{3} = \dfrac{75}{3}$

Results in: $1x = 25$

Or more conventionally: $x = 25$

[Solution]

To **check our answer**, we can substitute

25 for x in the original equation: $\dfrac{3}{15} = \dfrac{5}{25}$

The cross products become: $3 \cdot 25 = 15 \cdot 5$

Resulting in: $75 = 75$

Therefore, $x = 25$ is a solution to this proportional equation

Here is an example that has more complex rational expressions in the proportion,

Example 2.   Solve for x: $\dfrac{x+1}{x-3} = \dfrac{5}{6}$

Since cross products of a proportion are equal: $(x+1)(6) = (x-3)(5)$

Using the commutative property of multiplication: $(6)(x+1) = (5)(x-3)$

Distributing both sides of the equation: $6x + 6 = 5x - 15$

We now bring the "x" terms to the left side of the equation: $-5x \qquad\quad -5x$

Resulting in: $1x + 6 = -15$

We now isolate the "x" term by adding -6 to both sides of the equation: $-6 \qquad\quad -6$

Resulting in: $1x = -21$

Or more conventionally: $x = -21$

[Solution]

We will now look at some applications that use proportional equations.

Example 3.

Tom goes to the supermarket and purchases 4 - 12 oz. cans of red kidney beans for which he is charged $4.76. How much would he have to pay for 7 - 12 oz. cans?

We can **set up a proportional equation** to solve this problem.

Step 1. Define the variable for what we are trying to find.

We are asked to find the cost for 7 – 12 oz. cans of red kidney beans. Therefore
x = **the cost of 7 – 12 oz. cans of red kidney beans**

Step 2. Set up a proportion from the given information.

The ratios in this case are: $\dfrac{\text{number of cans}}{\text{cost for that number}}$

The **first ratio** given to us is: $\dfrac{4\,[\text{cans}]}{4.76\,[\text{cost}]}$

We can create the **second ratio** using

the variable that we have defined: $\dfrac{7\,[\text{cans}]}{x\,[\text{cost}]}$

We now use this information to set up the proportion: $\dfrac{4}{4.76} = \dfrac{7}{x}$

Since cross products of a proportion are equal: $4 \cdot x = 4.76 \cdot 7$

Multiplying: $4x = 33.32$

Dividing both sides of the equation by 4: $\dfrac{\cancel{4}x}{\cancel{4}} = \dfrac{33.32}{4}$

Results in: $x = 8.33$

Going back to our definition of the variable:

The cost of 7- 12 oz. cans of red kidney beans is $8.33
[Solution]

Example 4.

A robot that weighs 595 lbs. on the planet Earth would weigh 226 lbs. on Mars. Scientists are designing a new robot that will weigh 391 lbs. on Mars. What will it weigh on Earth? Round your answer to the nearest pound.

We can **set up a proportion** to solve this problem.

Step 1. Define the variable for what we are trying to find.

We are asked to find the weight of new robot on Earth. Therefore
x = **the weight of the new robot on Earth**

Step 2. Set up a proportion from the given information.

The ratios in this case are: $\dfrac{\text{weight on Earth}}{\text{weight on Mars}}$

The **first ratio** given to us is: $\dfrac{595\ [\text{Earth}]}{226\ [\text{Mars}]}$

We can create the second ratio using

the variable that we have defined: $\dfrac{x\ [\text{Earth}]}{391\ [\text{Mars}]}$

We now use this information to set up the proportion: $\dfrac{595}{226} = \dfrac{x}{391}$

Since cross products of a proportion are equal: $595 \cdot 391 = 226 \cdot x$

Multiplying: $232{,}645 = 226x$

Dividing both sides of the equation by 226: $\dfrac{232{,}645}{226} = \dfrac{\cancel{226}x}{\cancel{226}}$

Results in: $1029 \approx x$

[rounded to the nearest pound]

Going back to our definition of the variable:

The weight of the new robot on Earth is approximately 1029 lbs.

[Solution]

Example 5.

If three times a certain number is subtracted from the quotient of 7 and 4, the result is the number divided by 2. What is the number?

Solution.

First, we translate this problem into a mathematical equation and then use a proportion to solve it. We set up the equation as follows:

Step 1. Define the variable for what we are trying to find.

We are asked to find the number. Therefore
$$x = \textbf{the number}$$

Step 2. Translate the information given into mathematical terms.

3 times a certain number: $3x$

the quotient of 7 and 4: $\dfrac{7}{4}$

the number divided by 2: $\dfrac{x}{2}$

Step 3. Write an equation from the given information.

Be careful with subtraction, as the ***order*** of **what** is being subtracted from **what** is critical. In this case, $3x$ is being subtract **from** $\dfrac{7}{4}$

Therefore, $\dfrac{7}{4} - 3x$ represents the **left side of the equation.**

$$\left[\dfrac{7}{4} - 3x \text{ is } \underline{not} \text{ equivalent to } 3x - \dfrac{7}{4}. \text{ Subtraction is not commutative.}\right]$$

"the result is" in a sentence denotes the **"="** in an equation, so what follows **"the result is"** belongs on the **right side of the equation** $\left[\dfrac{x}{2}\right]$.

Therefore: $\dfrac{7}{4} - 3x = \dfrac{x}{2}$

If we use a proportional approach, we must treat the **entire** left side of the equation as a **unified** fraction in order to treat it as a proportion.

We can accomplish this by putting it all over 1:

$$\frac{\frac{7}{4}-3x}{1}=\frac{x}{2}$$

Now we can use the cross products concept:

$$\left(\frac{7}{4}-3x\right)\cdot 2 = 1\cdot x$$

Using the commutative property of multiplication on the left side of the equation:

$$2\left(\frac{7}{4}-3x\right)=1\cdot x$$

Distributing on the left side:

$$\frac{2}{1}\cdot\frac{7}{4}-2\cdot 3x = 1\cdot x$$

Factoring and canceling the fractions:

$$\frac{\cancel{2}}{1}\cdot\frac{7}{\cancel{2}\cdot 2}-2\cdot 3x = 1\cdot x$$

Multiplying:

$$\frac{7}{2}-6x = 1x$$

Isolating the x terms on the right side of the equation by adding $6x$ to both sides:

$$\underline{6x \quad\quad 6x}$$

Results in:

$$\frac{7}{2}\ =\ 7x$$

Again, creating a proportion by changing $7x$ to a fraction:

$$\frac{7}{2}\ =\ \frac{7x}{1}$$

Again, using the cross products concept:

$$7\cdot 1\ =\ 2\cdot 7x$$

Multiplying:

$$7\ =\ 14x$$

Dividing both sides of the equation by 14:

$$\frac{7}{14}=\frac{\cancel{14}x}{\cancel{14}}$$

Results in:

$$\frac{7}{14}=x$$

Reducing the fraction by factoring and canceling:

$$\frac{\cancel{7}\cdot 1}{\cancel{7}\cdot 2}\ =\ x$$

Results in:

$$\frac{1}{2}\ =\ x$$

Or more conventionally:

$$x=\frac{1}{2}$$

[solution]

It is important to note that to use the cross products concept for solving equations, we must treat the entire left and right side of the equations as unified fractions. If either or both are not unified, we can make them so by putting the entire expression over 1.

You also have the option to attack such a problem as we did in the previous topic: find a least common denominator and multiply every term in the equation by that quantity. The solution would be as follows:

$$\textbf{Solve for } x: \qquad \frac{7}{4} - 3x = \frac{x}{2}$$

As the LCD is 4 in this case, we will multiply both sides of

the equation by 4: $\qquad 4\left(\frac{7}{4} - 3x\right) = \frac{4}{1} \cdot \frac{x}{2}$

Distributing results in: $\qquad \frac{4}{1} \cdot \frac{7}{4} - 4 \cdot 3x = \frac{4}{1} \cdot \frac{x}{2}$

Factoring and canceling: $\qquad \frac{\cancel{4}}{1} \cdot \frac{7}{\cancel{4}} - 4 \cdot 3x = \frac{\cancel{2} \cdot 2}{1} \cdot \frac{x}{\cancel{2}}$

Multiplying: $\qquad \frac{7}{1} - 12x = \frac{2x}{1}$

We can now remove the "1" denominators: $\qquad 7 - 12x = 2x$

We move the x terms to the right side of the equation by adding $12x$ to each side:

$$\begin{array}{r} 12x \qquad 12x \\ \hline \end{array}$$

Resulting in: $\qquad 7 \qquad = 14x$

Dividing both sides of the equation by 14: $\qquad \frac{7}{14} = \frac{14x}{14}$

Results in: $\qquad \frac{7}{14} = 1x$

Factoring and canceling the left side of the equation: $\qquad \frac{\cancel{7} \cdot 1}{\cancel{7} \cdot 2} = 1x$

Results in: $\qquad \frac{1}{2} = 1x$

Or more conventionally: $\qquad x = \frac{1}{2}$

[Solution]

As can be seen, each method leads to the same solution.
In Algebra there is usually **more than one way to reach a solution**.

Concept Homework

1. How can we rewrite $\dfrac{p}{q} = \dfrac{r}{s}$ as an equation without fractions?

2. Are all of the following statements true? Explain why or why not.

    a.   If $\dfrac{p}{q} = \dfrac{r}{s}$ then $rq = sp$

    b.   If $\dfrac{p}{q} = \dfrac{r}{s}$ then $ps = rq$

    c.   If $\dfrac{p}{q} = \dfrac{r}{s}$ then $tps = trq$

3. Is $\dfrac{3}{y} + \dfrac{2}{x} = \dfrac{7}{z}$ a proportion? Explain why or why not.

4. Is $4 - x$ always equivalent to $x - 4$? Explain why or why not. Substitute a value for x to support your case.

Solutions: 1. $ps = qr$; 2. They are all true. 3. Only if $\dfrac{3}{y} + \dfrac{2}{x}$ is expressed as a unified fraction:; $\left(\dfrac{\frac{3}{y} + \frac{2}{x}}{1} \right)$ 4. No. Subtraction is not

commutative.

Exercises

Solve the following equations using proportions.

1. $\dfrac{6}{x} = \dfrac{2}{3}$ 2. $\dfrac{16}{x-2} = \dfrac{8}{x}$ 3. $\dfrac{9}{x+2} = \dfrac{3}{x-2}$ 4. $1 + \dfrac{9}{x} = \dfrac{7}{4}$

5. Jane begins an auto trip with a full tank of gas holding 16 gallons of gasoline. She stops to fill her tank which took 12 gallons. She checked her odometer and found that she had driven 276 miles on those 12 gallons. At this rate, how many miles can she drive on a full tank of gas?

6. A simple syrup is made by dissolving 2 cups of sugar in $\dfrac{2}{3}$ cups of boiling water. At this rate, how many cups of sugar are required for 2 cups of boiling water?

7. If twice a certain number is subtracted from the quotient of 5 and 6, the result is the number divided by 3. What is the number?

Solutions: **1.** $x = 9$; **2.** $x = -2$; **3.** $x = 4$; **4.** $x = 12$; **5.** 368 miles; **6.** 6 cups; **7.** $x = \dfrac{5}{14}$.

Topic 27 - Radical Expressions

The general form of a radical expression is: $\sqrt[a]{b}$ where b represents a number called the __radicand__ [the number inside the radical] and a represents a number called the __index__. The index indicates the **root** of radical.

When the **index** is "2", it is **not written**. For example $\sqrt[2]{5}$ is written without the index, 2. It is written $\sqrt{5}$. This is verbalized as **"the square root of 5"**. Like variables, radicals often have **implied** information. For example, $\sqrt{4}$ would have a the coefficient of the radical of 1 and an index of 2. If we wrote out all the implied information, it would look like $1 \cdot \sqrt[2]{4}$.

This is similar to **implied** information with variable expressions. For example, x implies $1 \cdot x^1$.

      Other Descriptive Examples: $\sqrt[3]{8}$ is verbalized as **"the cube root [or 3rd root] of 8"**
                                               $\sqrt[4]{16}$ is verbalized as **"the 4th root of 16"**

How do we convert square roots to "regular numbers"?

In the scope of this topic, we will **mostly** consider **square roots** $\left[\sqrt{}\right]$.

                [Those radicals with the implied index of 2]

We ask ourselves, **"What number broken down into *2 identical positive factors* will result in the radicand?" The answer to this question is the "regular" number value of the radical.**

    Example 1.   **What is the "regular" number value of:** $\sqrt{4}$?

      The **radicand** has a value of 4. Can 4 be broken down into two **identical positive factors**? Yes! $[2 \cdot 2 = 4]$. Therefore, since the radicand, 4, can be expressed as the equivalent value $2 \cdot 2$, $\sqrt{4}$ has
                                 a "regular" number value of 2: $\sqrt{4} \Rightarrow \sqrt{2 \cdot 2} = 2$
Since square roots are evaluated as positive numbers, we can express this in general form as $\sqrt{a^2} \Rightarrow \sqrt{a \cdot a} = |a|$. The absolute value shows us that the result must be positive.

If we want to express a negative square root, we would express it as $-\sqrt{a}$.

Example 2.   **What is the "regular" number value of:** $\sqrt{9} \Rightarrow \sqrt{3 \cdot 3} = 3$

Example 3.   **What is the "regular" number value of:** $\sqrt{49} \Rightarrow \sqrt{7 \cdot 7} = 7$

                       What is the "regular" number value of: $\sqrt{36} \Rightarrow \sqrt{6 \cdot 6} = 6$

Example 4.   **What is the "regular" number value of:** $-\sqrt{9} \Rightarrow -\sqrt{3 \cdot 3} = -3$

When square roots can be converted into exact values as in the above examples, we call them **perfect square radicals**. You should become familiar with some common perfect square roots :

$$\sqrt{1} \; [1 \cdot 1]; \; \sqrt{4} \; [2 \cdot 2]; \; \sqrt{9} \; [3 \cdot 3], \; \sqrt{16} \; [4 \cdot 4]; \; \sqrt{25} \; [5 \cdot 5]; \; \sqrt{36} \; [6 \cdot 6]; \; \sqrt{49} \; [7 \cdot 7]$$

We will be looking for these perfect square roots continually throughout this topic.

Radicands may contain **variables.** They are treated the same way as numbers.

Example 5.   **Simplify:** $\sqrt{x^2}$

     The radicand can be factored into two identical factors: $\sqrt{x^1 \cdot x^1}$

Notice that $x^1 \cdot x^1 = x^2$ **so** $\sqrt{x^2}$ is a **perfect square root.** Therefore,

               $\sqrt{x^2}$ is **equivalent to** x^1. So: $\sqrt{x^2} = |x|$

                                            [x **can only be**

positive.]

Example 6.   **Simplify:** $\sqrt{p^6}$

         We can factor p^6 into two identical factors: $\sqrt{p^3 \cdot p^3}$

     So $\sqrt{p^6}$ is a **perfect square root** $[p^3 \cdot p^3 = p^6]$**,** so: $\sqrt{p^6} = p^3$

     In fact, for any variable radicand with an **even exponent**, the square root can be evaluated as the variable with an exponent **half** as large as the exponent of the radicand .

Examples: $\sqrt{p^{10}} \Rightarrow \sqrt{p^5 \cdot p^5} = p^5$; $\sqrt{x^8} \Rightarrow \sqrt{x^4 \cdot x^4} = x^4$; $\sqrt{p^{18}} \Rightarrow \sqrt{p^9 p^9} = p^9$; etc.

Radicands will **not always** be perfect squares and will not always be equivalent to a number or variable without the radical symbol. We will now discuss such a case.

Example 7.

What is the *approximate* value of : $\sqrt{10}$

We ask ourselves, "What two positive identical factors multiplied together will result in 10? Upon some trial and error of various possibilities $[2 \cdot 2 = 4; \ 3 \cdot 3 = 9; \ 4 \cdot 4 = 16]$, we realize that there is no such positive identical factors that equal 10. So we then ask ourselves, "What **perfect square root** is closest to, but **less than** $\sqrt{10}$"? The closest one is $\sqrt{9} = 3$ $[3 \cdot 3 = 9]$

Then we ask ourselves, "What perfect square radical is closest to, but more than $\sqrt{10}$?" The closest is $\sqrt{16} = 4$ $[4 \cdot 4 = 16]$

Therefore, $\sqrt{10}$ has a "regular" number value somewhere between $\sqrt{9}$ [which is equivalent to 3] and $\sqrt{16}$ [which is equivalent to 4] but **we don't know** *exactly* **what it is**! Therefore, the **approximate** value of $\sqrt{10}$ **is between** 3 **and** 4. Using **inequality notation:** $3 < \sqrt{10} < 4$. In fact, $\sqrt{10}$ is an **irrational** number and we **cannot** determine its exact value.

Cube Roots $\left[\sqrt[3]{}\right]$ [The cube (third) root]

Although we won't be dealing with **cube roots** very often in this discussion, we should know how they are defined and we should be able to recognize the smaller "**perfect cube roots**".

How do we evaluate a perfect cube root $\left[\sqrt[3]{}\right]$?

We ask ourselves, "What **three** [since the index is 3] **identical factors** will result in the radicand?" This answer to this question will be the "**regular**" [not inside the radical] value of the radical. They are called **perfect cube radicals**.

Example 7.

Evaluate: $\sqrt[3]{8}$?

[In words we would say, "What is the cube root of 8?]

We can factor 8 into three **identical factors** $[2 \cdot 2 \cdot 2]$. So: $\sqrt[3]{8} = \sqrt[3]{2 \cdot 2 \cdot 2}$

Therefore: $\sqrt[3]{8} = 2$

[Solution]

Example 8.   **Evaluate:** $\sqrt[3]{27}$

We can factor 27 into three **identical factors** [3·3·3]. So: $\sqrt[3]{27} = \sqrt[3]{3 \cdot 3 \cdot 3}$

Therefore: $\sqrt[3]{27} \;=\; 3$

[Solution]

Other Examples of Perfect Cub Radicals:

$$\sqrt[3]{64} \Rightarrow \sqrt[3]{4 \cdot 4 \cdot 4} = 4 ; \;\; \sqrt[3]{125} \Rightarrow \sqrt[3]{5 \cdot 5 \cdot 5} = 5 ; \;\; \sqrt[3]{p^3} \Rightarrow \sqrt[3]{p^1 \cdot p^1 \cdot p^1} = p^1 \text{ or just } p \,;$$

$$\sqrt[3]{x^6} = \sqrt[3]{x^2 \cdot x^2 \cdot x^2} = x^2 \text{ , etc.}$$

Operations with Radical Expressions $[+, -, \times, \div, \text{exponentials}]$

Since radical expressions are special types of numbers, just as **variable** expressions [polynomials] and **exponential** expressions, we have special rules for **multiplication, division, addition and subtraction. Radical expressions** operations are similar to **polynomial operations.** Therefore, if you are unsure of how to operate with them, ask yourself, **"What would I do if the radical was a variable?"**

1. Simplifying and Multiplication of Radicals: The Product Rule of Radicals

As long as the **indexes are the same:**

> When radicals are multiplied, the radicands can be put under one radical as factors, and, a radicand that has been factored, can be broken into separate radicals.

For square roots, this can be expressed mathematically as:

$$\boxed{\sqrt{a} \cdot \sqrt{b} = \sqrt{a \cdot b} \quad \text{or:} \quad \sqrt{a \cdot b} = \sqrt{a} \cdot \sqrt{b}}$$

Some Examples: $\sqrt{2} \cdot \sqrt{3} \Rightarrow \sqrt{2 \cdot 3}$ or $\sqrt{6}$; $\sqrt{10} = \sqrt{5 \cdot 2}$ or $\sqrt{5} \cdot \sqrt{2}$

Using the "Product Rule" to Simplify Radicals

We can often simplify radicals that are not perfect squares by **taking out from the radical any perfect square factors** contained in the radicand and expressing them as regular number multipliers [coefficients] of what is left in the radical.

Example 1.

Simplify: $\sqrt{8}$

There is no perfect square root of 8 [no two identical factors multiplied together result in 8]. **However** 8 has a **perfect square factor** [4]. We can **factor the radicand:** $\sqrt{4 \cdot 2}$

and use the **Product Rule,** $\left[\sqrt{a \cdot b} = \sqrt{a} \cdot \sqrt{b}\right]$, **to create two radicals:** $\sqrt{4} \cdot \sqrt{2}$

As $\sqrt{4}$ has is equivalent to 2 : $2 \cdot \sqrt{2}$

2 becomes the **coefficient** [multiplier] of the radical $\sqrt{2}$: $2\sqrt{2}$

[simplified form]

Example 2.

Simplify: $-4\sqrt{27}$

$\sqrt{27}$ has a "perfect square" factor [9]: $-4\sqrt{9 \cdot 3}$

Using the **Product Rule:** $-4\sqrt{9} \cdot \sqrt{3}$

Since $\sqrt{9} = 3$: $-4 \cdot 3 \cdot \sqrt{3}$

Since -4 and 3 are numbers outside the radical, we can multiply them: $-12\sqrt{3}$

[simplified form]

Example 3.

Simplify: $\dfrac{\sqrt{24}}{6}$

Since $\sqrt{24}$ has a "perfect square" factor, 4 : $\dfrac{\sqrt{4 \cdot 6}}{6}$

Using the **Product Rule:** $\dfrac{\sqrt{4} \cdot \sqrt{6}}{6}$

Since $\sqrt{4} = 2$: $\dfrac{2\sqrt{6}}{6}$

We **cannot** "cancel" the $\sqrt{6}$ in the numerator [a radicand] with the 6 in the denominator, because they are different **"types"** of numbers. However, the coefficient, 2 , in the numerator and the denominator, 6 are both **like** kinds of numbers acting as factors in the radical

expression. Therefore they **can** be reduced: $\dfrac{\cancel{2}\sqrt{6}}{\cancel{2} \cdot 3}$

Resulting in: $\dfrac{1\sqrt{6}}{1 \cdot 3}$

Multiplying by 1, has no effect on the value, so it can be omitted: $\dfrac{\sqrt{6}}{3}$

[completely simplified]

The same simplification techniques can be used when radicals contain variables. Here is an example.

Example 4.   Simplify: $\sqrt{x^5}$

We know that any radicand with an even exponent is a perfect square root and can be removed from the radical. We can rewrite $\sqrt{x^5}$ as: $\sqrt{x^4 \cdot x^1}$

Now we can separate this radical into two components **[Product Rule]** : $\sqrt{x^4} \cdot \sqrt{x^1}$

Since $\sqrt{x^4} \Rightarrow \sqrt{x^2 \cdot x^2}$ is a perfect square root it can be expressed as: $x^2 \cdot \sqrt{x^1}$

and the simplification is conventionally expressed as: $x^2\sqrt{x}$

[completely simplified]

Other Examples:

a. $\sqrt{32y^7} \Rightarrow \sqrt{16 \cdot 2 \cdot y^6 \cdot y^1} \Rightarrow \quad \sqrt{16} \cdot \sqrt{y^6} \cdot \sqrt{2 \cdot y^1} \Rightarrow \quad 4 \cdot y^3 \cdot \sqrt{2y} = \quad 4y^3\sqrt{2y}$

[product rule / commutative property]   [completely simplified]

b. $\sqrt{40y^{11}} \Rightarrow \sqrt{4 \cdot 10 \cdot y^{10} \cdot y^1} \Rightarrow \quad \sqrt{4} \cdot \sqrt{y^{10}} \cdot \sqrt{10 \cdot y^1} \Rightarrow \quad 2 \cdot y^5 \cdot \sqrt{10y^1} = \quad 2y^5\sqrt{10y}$

[product rule / commutative property]   [completely simplified]

Multiplying radicals is similar to multiplying polynomials as can be seen in the next example.

Example 5.   Multiply and simplify: $3\sqrt{6} \cdot 4\sqrt{3}$

Since all the operations are multiplication, this is a **series of factors**: $3 \cdot \sqrt{6} \cdot 4 \cdot \sqrt{3}$
which can be re-arranged so that the coefficients are together and the

radicals are together **[the commutative property of multiplication in action]**: $3 \cdot 4 \cdot \sqrt{6} \cdot \sqrt{3}$

[Notice that this is similar to $3x \cdot 4y = 3 \cdot 4 \cdot x \cdot y$ when operating with polynomials.]

The coefficients are multiplied and, using the **product rule**, the

radicals are combined: $12\sqrt{6 \cdot 3}$

As there are no perfect square factors in 6 and 3 we will multiply them

and see what develops: $12\sqrt{18}$

$\sqrt{18}$ can be simplified since it has a perfect square factor of 9: $12\sqrt{9 \cdot 2}$

which becomes a separate radical **[The Product Rule]**: $12\sqrt{9} \cdot \sqrt{2}$

Since $\sqrt{9} = 3$: $12 \cdot 3\sqrt{2}$

multiplying the coefficients results in: $36\sqrt{2}$

[Solution]

Now the result of the multiplication is simplified.

Here is an example that contains variables in the radical:

Example 6.   **Multiply and Simplify:** $4\sqrt{x}\cdot 3\sqrt{x}$

This multiplication can be expressed as a series of factors: $4\cdot\sqrt{x}\cdot 3\cdot\sqrt{x}$

The commutative property of multiplication allows us to
re-arrange the order of the factors: $4\cdot 3\cdot\sqrt{x}\cdot\sqrt{x}$

Multiplying and using the **product rule:** $12\cdot\sqrt{x\cdot x}$

Since the radicand is a perfect square, its equivalent can come out
of the radical: $12x$
[Solution]

2. Division and the Quotient Rule of Radicals

Division of radicals occurs when the radicand is expressed as a fraction.

Example 1.   **Simplify:** $\sqrt{\dfrac{36}{4}}$

Since a fraction represents a division operation we can divide 36 by 4: $\sqrt{9}$

As $\sqrt{9}$ is a perfect square root $[3\cdot 3=9]$, we can express it as a
"regular number: 3
[completely simplified]

Example 2.   **Simplify:** $\dfrac{\sqrt{36}}{\sqrt{4}}$

As each radical in the fraction is a perfect square root, it is equivalent to: $\dfrac{6}{2}$

Since a fraction represents a division operation we can divide 6 by 2: 3
[completely simplified]

Note that both Example 1 and Example 2 yields the **same results**. Therefore,

$$\sqrt{\frac{36}{4}} = \frac{\sqrt{36}}{\sqrt{4}}.$$

If we generalize this result, we arrive at the **Quotient Rule of Radicals**:

$$\sqrt{\frac{a}{b}} = \frac{\sqrt{a}}{\sqrt{b}} \quad\text{or}\quad \frac{\sqrt{a}}{\sqrt{b}} = \sqrt{\frac{a}{b}} \qquad [b\neq 0]$$

244

Here are some examples of how we can utilize this rule to simplify radicals that involve division.

Example 3.   **Simplify using the quotient rule:** $\sqrt{\dfrac{12}{25}}$

Using the quotient rule $\left[\sqrt{\dfrac{a}{b}} = \dfrac{\sqrt{a}}{\sqrt{b}}\right]$ the radical can be expressed as : $\dfrac{\sqrt{12}}{\sqrt{25}}$

Factoring out the perfect square factor, 4, in the numerator: $\dfrac{\sqrt{4\cdot3}}{\sqrt{25}}$

Using the product rule in the numerator: $\dfrac{\sqrt{4}\cdot\sqrt{3}}{\sqrt{25}}$

Expressing the perfect square roots as "regular numbers": $\dfrac{2\cdot\sqrt{3}}{5}$

Or more conventionally: $\dfrac{2\sqrt{3}}{5}$

[completely simplified]

We will now give an example with variables in the radical.

Example 4.   **Simplify using the quotient rule:** $\sqrt{\dfrac{50x^7}{9y^6}}$

Since there is no "canceling" possible in the radicand, we will start by using the quotient rule to create two different radicals: $\dfrac{\sqrt{50x^7}}{\sqrt{9y^6}}$

Factoring out the perfect square factors in the numerator: $\dfrac{\sqrt{25\cdot2\cdot x^6\cdot x^1}}{\sqrt{9y^6}}$

We re-write the fraction using the product rule and commutative property: $\dfrac{\sqrt{25}\cdot\sqrt{x^6}\cdot\sqrt{2x^1}}{\sqrt{9}\cdot\sqrt{y^6}}$

Expressing the perfect square radicals as "regular numbers": $\dfrac{5x^3\sqrt{2x}}{3y^3}$

[completely simplified]

3. Addition and Subtraction of Radicals

As we have seen, radicals are different **types** of numbers, that operate in ways similar to variables in polynomial expressions. Under addition, they have to be treated in the same way as variables in polynomial expressions. For example, the number, 2 [outside the radical], **cannot** be added to the radicand, 3, in $\sqrt{3}$. Therefore we cannot simplify $2+\sqrt{3}$ [This is similar to "like terms" with variables: $2+x$ can't be simplified]. Addition and subtraction rules are similar to those of polynomials. Only "like radicals" can be added or subtracted. **The radical parts of radical expressions must be identical to be added or subtracted. The coefficients are the only parts that are added or subtracted.**

Example 1. **Simplify:** $3\sqrt{2}+2\sqrt{2}$

Since the radical parts of these expressions are **identical** [they have the same radicands and indexes], the terms can be added. This is done by adding the coefficients and keeping the radical parts unchanged: $5\sqrt{2}$

[Notice that this is similar to working with polynomials: $3x+2x=5x$] [Solution]

Example 2. **Simplify:** $\sqrt{3}+\sqrt{3}$

This problem becomes clearer when we add the "1" coefficients: $1\sqrt{3}+1\sqrt{3}$

Adding the coefficients and keeping the radicals the same results in: $2\sqrt{3}$

[This is similar to working with polynomial expression: $x+x \Rightarrow 1x+1x=2x$] [Solution]

Example 3. **Simplify:** $2\sqrt{5}+3\sqrt{7}$

Since the radical parts of the expressions are **not** identical [the radicands are different.], these expressions **cannot be simplified.**

[This is similar to the way we deal with the polynomial expression, $2x+3y$ which cannot be simplified!]

Example 4. **Simplify:** $\left(3+\sqrt{5}\right)+\left(4+2\sqrt{5}\right)$

The parentheses are removed since there is no subtraction or multiplication involved between the parentheses. By adding a "1" coefficient to $\sqrt{5}$, the problem becomes clearer: $3+1\sqrt{5}+4+2\sqrt{5}$

We can only add like terms! We add 3 and 4 [numbers in the expression that are not involved with radicals]: $7+1\sqrt{5}+2\sqrt{5}$

$1\sqrt{5}$ and $2\sqrt{5}$ have identical radicals so we can add them also: $7+3\sqrt{5}$

[Solution]

[In this last step, notice that we only added the coefficients and kept the radicals the same. This is similar to adding variable expressions like $1x + 2x = 3x$. The variable part stays the same. Only the coefficients are added!]

Example 5.

Simplify: $\left(4 - \sqrt{3}\right) - \left(3 + 5\sqrt{3}\right)$

Adding a "1"coefficient to $\sqrt{3}$ to make the problem clearer: $\left(4 - 1\sqrt{3}\right) - \left(3 + 5\sqrt{3}\right)$

Just as we do with polynomials, we change the subtraction between the expressions to addition by reversing the signs of **every term** in the parenthesis following it: $\left(4 - 1\sqrt{3}\right) + \left(-3 - 5\sqrt{3}\right)$

We can now remove the parentheses: $4 - 1\sqrt{3} + \left(-3\right) - 5\sqrt{3}$

We can only add like terms! We add 4 and -3 [**numbers in the expression that are not involved with radicals**]: $1 - 1\sqrt{3} - 5\sqrt{3}$

$-1\sqrt{3}$ and $-5\sqrt{3}$ have identical radicals so we can add them also: $1 - 6\sqrt{3}$

[Solution]

[Notice that we are following the same rules as if the radical expressions were polynomial expressions. This problem was approached in the same manner as we would approach $\left(4 - x\right) - \left(3 + 5x\right)$. The result is $1 - 6x$.]

Example 6.

Add and Simplify: $\sqrt{27} + \sqrt{75}$

At first glance, we see that **the radicands are not identical** and might assume that they cannot be combined. But **these radicands have "perfect square" factors!** This fact might change the situation. Factoring the radicands results in: $\sqrt{9 \cdot 3} + \sqrt{25 \cdot 3}$

Using the **product rule**, the **perfect square factors** can be treated as separate radicals: $\sqrt{9} \cdot \sqrt{3} + \sqrt{25} \cdot \sqrt{3}$

Evaluating the perfect square radicals results in: $3 \cdot \sqrt{3} + 5 \cdot \sqrt{3}$

Or just: $3\sqrt{3} + 5\sqrt{3}$

We **now have** identical radicands that can be added: $8\sqrt{3}$

[Solution]

3. Distributing and "FOILing" Radical Expressions

We distribute with radical expressions **just as we would with polynomials!**

Example 1.

Distribute and Simplify: $\sqrt{8}\left(2\sqrt{3}+\sqrt{2}\right)$

First we multiply each term within the parenthesis by $\sqrt{8}$:

$\sqrt{8}\cdot 2\sqrt{3}+\sqrt{8}\cdot\sqrt{2}$

We use the commutative property to rearrange the radicals in the first term: $2\sqrt{3}\cdot\sqrt{8}+\sqrt{8}\cdot\sqrt{2}$

We use the **Product Rule** $\left[\sqrt{a}\cdot\sqrt{b}=\sqrt{a\cdot b}\right]$, to combine the radicals: $2\sqrt{8\cdot 3}+\sqrt{8\cdot 2}$

and multiply the factors of each radicand: $2\sqrt{24}+\sqrt{16}$

Now we look for **perfect square** factors for the radicand, 24 , and convert $\sqrt{16}$ to its "regular" number equivalent, 4 , resulting in: $2\sqrt{4\cdot 6}+4$

Using the **Product Rule** results in: $2\sqrt{4}\cdot\sqrt{6}+4$

Since $\sqrt{4}=2$ to a "regular" number we now have: $2\cdot 2\cdot\sqrt{6}+4$

Multiplying the coefficients results in: $4\sqrt{6}+4$

[**Notice that we cannot combine the** $4's$ **since one is a coefficient and the other is a** *constant*, **just as we couldn't combine** $4x+4$ **.**]

Example 2.

Multiply and Simplify: $\left(\sqrt{5}-4\right)\left(\sqrt{3}+2\right)$

If we look carefully at the makeup of this problem, we see that it is similar to multiplying two binomials. Each factor is made of two different **kinds** of numbers: a regular number and a radical. This multiplication is handled in the same way as we would with binomials: $(x-4)(y+2)$. Therefore, we will use a FOIL approach.

We start by putting in "1" coefficients: $\left(1\sqrt{5}-4\right)\left(1\sqrt{3}+2\right)$

Multiplying **F**irst, **O**utside, **I**nside and **L**ast terms in the expressions:

First terms: $1\sqrt{5}\cdot 1\sqrt{3}\Rightarrow 1\cdot 1\cdot\sqrt{5}\cdot\sqrt{3}=1\sqrt{15}$

Outside terms: $2\cdot 1\sqrt{5}=2\sqrt{5}$

Inside terms: $-4\cdot 1\sqrt{3}=-4\sqrt{3}$

Last terms: $-4\cdot 2=-8$

Adding these products together: $1\sqrt{15}+2\sqrt{5}-4\sqrt{3}-8$

$1\sqrt{15}$, $2\sqrt{5}$ and $4\sqrt{3}$ are not "like" radicals **[the radicands are not identical]**. Also, they do not have any perfect square factors. So

the most simplified form of the expression is: $\sqrt{15} + 2\sqrt{5} - 4\sqrt{3} - 8$

[Solution]

Example 3.
Multiply and Simplify: $\left(\sqrt{6}+2\right)\left(\sqrt{6}-2\right)$

Upon close examination, we see that these factors look like **conjugates** [similar to $(x+2)(x-2)$].

Just like with binomial conjugates, when we FOIL, the middle terms will be **opposites** and sum to zero.

Adding "1" coefficients: $\left(1\sqrt{6}+2\right)\left(1\sqrt{6}-2\right)$

Performing the multiplication

First terms: $1\sqrt{6} \cdot 1\sqrt{6} \Rightarrow 1\sqrt{6 \cdot 6} = 1 \cdot 6$

Outside terms: $-2 \cdot 1\sqrt{6} = -2\sqrt{6}$

Inside terms: $2 \cdot 1\sqrt{6} = 2\sqrt{6}$

Last terms: $2 \cdot (-2) = -4$

Adding the results: $6 - 2\sqrt{6} + 2\sqrt{6} - 4$

The middle terms are **opposites** and **add to zero**: $6 + 0 - 4$

Resulting in: 2 [solution]

Whenever, we multiply "conjugate" radical expressions, the results **will not contain a radical!**

4. Exponentials : Squaring a square root
Evaluate: $\left(\sqrt{x}\right)^2$

The rules of exponents tell us that the base of this exponential is $\sqrt{x}$. The squaring process multiplies the base by itself.

Therefore, $\left(\sqrt{x}\right)^2 = \sqrt{x} \cdot \sqrt{x}$. So, $\left(\sqrt{x}\right)^2 = \sqrt{x} \cdot \sqrt{x}$

The **Product Rule,** $\sqrt{a} \cdot \sqrt{a} = \sqrt{a \cdot a}$ allows us to say that $\sqrt{x} \cdot \sqrt{x} = \sqrt{x \cdot x}$

A perfect square radical has identical factors, so, $\sqrt{x \cdot x} = x$

[Solution]

This will be true for any real value of x.

More examples:

a. $\left(\sqrt{2}\right)^2 \Rightarrow 2$ b. $\left(\sqrt{25}\right)^2 \Rightarrow 25$ c. $\left(\sqrt{(x+2)}\right)^2 = (x+2)$

Since $\left(\sqrt{x}\right)^2$ is equivalent to x and $\sqrt{x^2}$ is equivalent to x , then $\left(\sqrt{x}\right)^2 = \sqrt{x^2}$.

Concept Homework

As an assessment of your understanding of the concepts set forth in this section, answer the following questions. If necessary, review the material to help you arrive at the correct conclusions. With true or false statements, if the statement is true, make up an example that supports the statement without using examples given in the material. If a statement is false, correct the wording to make it true. Then make up an example that supports the statement without using examples given in the material.

<u>True of False</u>

1a. The index of $\sqrt{3}$ is 3.

   b. $\sqrt[3]{125} = 25$

   c. $\sqrt{27}$ can be evaluated exactly.

   d. $\sqrt{x^6} = x^3$

   e. $\dfrac{2\sqrt{3}}{3} \Rightarrow 2\sqrt{1} \Rightarrow 2 \cdot 1 = 2.$

   f. In order to **multiply** radical expressions, the radical parts must be identical. We multiply the coefficients and leave the radical parts as they were originally.

   g. In order to **add or subtract** radical expressions, the radical parts must be identical. We add or subtract the coefficients and leave the radical parts as they were originally. This is similar to adding or subtracting polynomials.

   h. The rules for operating $\left(+,-,\times,\div\right)$ on radical expressions are similar to those involving polynomials. Radicals behave just as variables for these operations.

   i. When multiplying radicals, both the coefficients of the radicals are multiplied and the radicands are multiplied. The radicands may combined as factors in one radical if they have the same indexes.

   j. The result of multiplying $\left(2\sqrt{5}+7\right)\left(2\sqrt{5}-7\right)$], will not contain a radical.

2. What 3 identical factors are there of x^9 ?

3. Evaluate $\sqrt[3]{x^9}$

4. What is the coefficient of $\sqrt{5}$?

5. What is the index of $\sqrt{5}$?

6. Is there an exact number value of $\sqrt{5}$? What whole numbers does it lie between?

7. Between what two whole numbers does $\sqrt{23}$ lie?

Solutions. **1a.** False; **b.** False; **c.** False; **d.** True ; **e.** False ; **f.** False ; **g.** True ; **h.** True ; **i.** True ; **j.** True ; **2.** $x^3 \cdot x^3 \cdot x^3$; **3.** x^3 ; **4.** 1 ; **5.** 2 ; **6.** No; 2 and 3 . **7** . Between 4 and 5.

250

Exercises

Simplify:

1. $\sqrt{24}$; **2.** $-\sqrt{27}$; **3.** $\sqrt{12y^3}$; **4.** $\sqrt{98x^5}$; **5.** $\sqrt{20x^2y^3}$ **6.** $\sqrt[3]{27y^3}$;

Perform the indicated operation and simplify:

7. $\sqrt{3}+2\sqrt{3}$; **8.** $2\sqrt{3}+\sqrt{27}$; **9.** $\sqrt{8}+\sqrt{18}$; **10.** $\sqrt{4a}+\sqrt{9a}$;

11. $\left(5-\sqrt{6}\right)-\left(2+3\sqrt{6}\right)$; **12.** $4\left(\sqrt{2}-\sqrt{3}\right)$; **13.** $2+2\sqrt{3}$; **14.** $\sqrt{3}\cdot\sqrt{6}$;

15. $\sqrt{3a}\cdot\sqrt{27a^2}$; **16.** $\sqrt{3}\left(2-\sqrt{15}\right)$; **17.** $\dfrac{\sqrt{8}}{\sqrt{2y}}$; **18.** $\left(3\sqrt{2}-4\sqrt{3}\right)\left(\sqrt{2}-3\sqrt{3}\right)$;

19. $\sqrt{\dfrac{27}{3}}$; **20.** $\sqrt{\dfrac{8a^2b^3}{3x^3y^2}}$; **21.** $\dfrac{\sqrt{24x^3y^4}}{\sqrt{2xy}}$; **22.** $\left(\sqrt{2x+3}\right)^2$ **23.** $\left(2+3\sqrt{5}\right)\left(2-3\sqrt{5}\right)$

Solutions: **1.** $2\sqrt{6}$; **2.** $-3\sqrt{3}$; **3.** $2y\sqrt{3y}$; **4.** $7x^2\sqrt{2x}$; **5.** $2xy\sqrt{5y}$; **6.** $3y$; **7.** $3\sqrt{3}$; **8.** $5\sqrt{3}$; **9.** $5\sqrt{2}$; **10.** Can't be added;

11. $3-4\sqrt{6}$; **12.** $4\sqrt{2}-4\sqrt{3}$; **13.** Can't be added; **14.** $3\sqrt{2}$; **15.** $9a\sqrt{a}$; **16.** $2\sqrt{3}-3\sqrt{5}$; **17.** $\dfrac{2\sqrt{2}}{\sqrt{2y}}$ or $2\sqrt{\dfrac{1}{y}}$;

18. $42-13\sqrt{6}$; **19.** 3 ; **20.** $\dfrac{2ab\sqrt{2b}}{xy\sqrt{3x}}$; **21.** $2xy\sqrt{3y}$; **22.** $2x+3$; **23.** -41 .

Topic 28 - Problem Solving: Formulas, Percent & Mixture Problems

When attempting to apply the algebraic principles to solving concrete problems [sometimes referred to as "word problems"], it is helpful to develop a general strategy. A recommended strategy has the following guidelines:

- "Word Problems" usually contain a lot of critical information in a very small space. Many times it is hard to digest all the information and how it is related in one reading of the problem. Take the time to **re-read the problem as many times as necessary**, so that you better understand what information is given and what you are being asked to solve.
- It is helpful to be able to determine if the problem falls into a **category** to which you have some familiarity. If you can apply a framework previously used, it can get you started on the right track towards a solution.
- In most cases, the question that you are being asked should lead to your **defining a variable or variables** that will eventually lead to an **equation** [or equations] that will solve the problem.
- We should always look at a solution and see whether it passes the "common sense" test. If the solution is outlandish, we have probably made an error either in developing an equation or in our calculations and should revisit the problem.

1. Isolating a Variable in Formulas

This is probably a familiar formula for finding the **Area of a rectangle**:

Area = Length times Width, or written in variables:
$$A = lw$$

With the formula in this format, if we are given the length and width of a rectangle, we can use this formula to find its area.

Suppose we wanted to find the length, given the area and width. We could <u>isolate</u> l in the formula by getting it alone on one side of the equation. We call this manipulation **solving for l in terms of A and w**. In effect, we are **isolating** the variable, l.

Example 1.   Given $A = lw$, solve for l **in terms of** A **and** w.

We start with the formula as given: $A = lw$

We can isolate l by dividing both sides of the equation by w : $\dfrac{A}{w} = \dfrac{lw}{w}$

Since $\dfrac{w}{w} = 1$ **[on the right side of the equation]**, they cancel each other leaving: $\dfrac{A}{w} = l$

Interchanging the left side with the right: $l = \dfrac{A}{w}$

We have accomplished our task of **solving for** l **in terms of** A *and* w.

Example 2.   The distance formula **[Distance = Rate × Time]** is symbolized as $D = rt$
Solve this equation for t **in terms of** D **and** r .

We are being asked to isolate the variable, t in the formula. We can

accomplish this by dividing both sides of the equation by r: $\dfrac{D}{r} = \dfrac{rt}{r}$

Since $\dfrac{r}{r} = 1$ **[on the right side of the equation]**, they cancel each other leaving: $\dfrac{D}{r} = t$

Interchanging the left side with the right: $t = \dfrac{D}{r}$

[Solution]

Example 3.

The perimeter of a rectangle is the distance around it. Since the lengths and widths of a rectangle are equal, the perimeter can be determined by adding twice the length and twice the width. This is represented by the formula $P = 2l + 2w$. Find w in terms of P and l.

We start by writing the formula as given: $P = 2l + 2w$

We need to isolate w. The first step is to remove the "$2l$" term from the right side of the equation by adding its opposite, $-2l$, to

both sides:

$$P = 2l + 2w$$
$$\underline{\quad -2l \qquad\qquad -2l \quad}$$

Resulting in: $\quad P - 2l = \quad 2w$

Dividing **all the terms** on both sides of the equation by 2: $\quad \dfrac{P - 2l}{2} = \dfrac{2w}{2}$

On the right side of the equation, the "2's" cancel : $\quad \dfrac{P - 2l}{2} = \dfrac{1w}{1}$

Or just: $\quad \dfrac{P - 2l}{2} = w$

Interchanging the right and left sides of the equation: $\quad w = \dfrac{P - 2l}{2}$

[Solution]

We don't necessarily have to isolate the required variable to solve for it in an equation. This next problem is an example:

Example 4.

A family drives from Cincinnati, Ohio to Rapid City, South Dakota, a distance of 1200 miles. They average a rate of 50 miles per hour. How much time did they spend driving?

We recognize this as falling into the **category** of a Distance, Rate and Time problem. We know that the formula, $D = rt$ will be involved.

So we will start by stating the formula: $\qquad D = rt$

We are asked to find t ["How much *time* did they spend driving?"]. We are given values for D [1200 miles] and r [50 miles per hour], so we substitute these values into the equation: $\qquad 1200 = 50t$

We now create a coefficient of 1 for t by dividing both sides of

the equation by 50: $\qquad \dfrac{1200}{50} = \dfrac{50t}{50}$

Resulting in: $\qquad 24 = 1t$

Or more simply: $\qquad 24 = t$

The time was stated in hours so the solution is:

"The family spent 24 hours driving."

Notice that it was not necessary to isolate t in the formula before solving the problem. We should ask ourselves if our solution "makes sense". It seems probable that it would take 24 hours to drive 1200 miles at 50 miles per hour. If our solution was 2400 hours, it would not pass the "common sense" test.

Example 5.

The length of a rectangle is one more meter than 4 times its width. Find the dimensions of the the rectangle if the perimeter is 52 meters.

We recognize this as a **perimeter** problem and will employ the formula: $P = 2l + 2w$
The problem supplies us with the perimeter, so we will substitute that
information for P in the formula: $52 = 2l + 2w$

We are asked to find l and w. We have **two** variables left in the formula. However, the problem gives us another equation dealing with l and W. It tells us the the length is 1 more than 4 times the width. Translating this into an equation we have: $l = 1 + 4w$

We now have a **system of two equations** with the same two
variables, l and W that can be solved: (1) $l = 1 + 4w$
(2) $52 = 2l + 2w$

Using the **substitution method**, we will substitute (1) into (2):   (1) $l = \underline{1 + 4w}$

(2) $52 = 2l + 2w$

Resulting in an equation with only one variable $[w]$: $52 = 2(1 + 4w) + 2w$

We can proceed with solving the equation for w. We will start
by distributing on the right side of the equation: $52 = 2 + 8w + 2w$
Then combining like terms: $52 = 2 + 10w$
We now isolate the w variable by adding -2 to both sides of the
equation: $\underline{-2 \quad -2}$
resulting in: $50 = 10w$
Dividing both sides of the equation by 10: $\dfrac{50}{10} = \dfrac{10w}{10}$
resulting in: $5 = w$

$$\text{or:} \quad w = 5$$

We have found a value for w [which is 5] and can use that in the

$$52 = 2l + 2w \text{ [Equation (1)]} \text{ to find a value for } l: \quad 52 = 2l + 2(5)$$

$$\text{Perform the multiplication:} \quad 52 = 2l + 10$$

$$\text{Isolate } l \text{ by add } -10 \text{ to both sides of the equation :} \quad \underline{-10 \qquad -10}$$

$$\text{Results in:} \quad 42 = 2l$$

$$\text{Dividing both sides of the equation by 2:} \quad \frac{42}{2} = \frac{2l}{2}$$

$$\text{Results in:} \quad 21 = l$$

[Note: We could have substituted $w=5$ into equation (2) $[l = 1 + 4w]$ and obtained the same result.]

So we have established that **the width of the rectangle is 5 meters** and the **length of the rectangle is 21 meters.**

Is this a reasonable solution? If we substitute 5 and 21 into both equations, we will see that they would both be true. Therefore our solution is both reasonable and correct.

2. Percent Problems

Finding percents usually fits into the following equation:

A partial amount is a percentage of the base (the whole amount)

[In translating words to mathematical symbols, "is" translates to "=", and "of" translates to multiplication.]

Translated into an equation this becomes:

$$p \text{ [partial amount]} = \% \cdot b \text{ [base or whole amount]}$$

Example 6.   The number 22 is what percent of 40?

We see that this is a percent problem so we will use the formula $\quad p = \% \cdot b$

We are asked to find the percent, so we will define the x as the percent. So now the formula looks like: $\quad p = x\,b$

We are told that p [partial or smaller amount] is 22 and b [whole or larger amount] is 40. Substituting these values into the formula:

$$22 = x \cdot 40$$

Using the commutative property this becomes:

$$22 = 40x$$

Dividing both sides of the equation by 40:

$$\frac{22}{40} = \frac{40x}{40}$$

results in:

$$\frac{22}{40} = x$$

We are interested in this number as a percent. First we will change $\frac{22}{40}$ to a decimal be dividing the numerator by the denominator:

$$.55 = x$$

Now we change the decimal to a percent by moving the decimal point two places to the right:

$$55\% = x$$

Stating the solution in terms of the original problem:

22 is 55% of 40

This solution passes the "common sense" test as 22 is slightly more than half of 40. We can verify that this is a correct solution since if we multiply .55 by 40, the result will be 22.

Example 7.   The number 150 is 40% of what number?

We see that this is a percent problem so we will use the formula

$$p = \% \cdot b$$

If 150 is 40% of some other number, 150 must be p [partial amount] of a whole [larger] amount b . Substituting these amounts into the formula will make it:

$$150 = 40.\% \cdot b$$

Changing the percent to a decimal by moving the decimal point two places to the right:

$$150 = .40b$$

Dividing both sides of the equation by .40:

$$\frac{150}{.40} = \frac{.40b}{.40}$$

results in:

$$375 = b$$

Stating the solution in terms of the original problem:

150 is 40% of 375

We knew that the solution had to be larger than 150 since 150 is only 40% of the solution. Our answer is larger than 150. We can verify that this is the correct solution by multiplying 375 by .40. The result will be 150.

Example 8.

A surfboard, originally purchased for $400, was sold on eBay at a discount of 40%. What is the eBay price and what is the discount?

As this is a percent problem we will use the formula : $\quad p \; = \; \% \; \cdot \; b$

The original price, b, is given as $400 [base or entire amount] and p represents the eBay price that is discounted by 40%. This means that p is 60% of b [100%-40%=60%] since it has been discounted by 40%. Substituting these values into the

formula results in: $\quad p = 60\% \cdot 400$

Changing the percent to a decimal: $\quad p = .60 \cdot 400$

Doing the multiplication: $\quad p = 240$

Therefore, the eBay price is $240.

The discount is the difference between the original purchase price and the eBay price. $400 - $240 = $160.

Therefore the discount is $160

Do these solutions make sense?

The discounted price is less than the original price which is as it should be. 40% of $400 [.40x400] is $160, so the **discount is correct**. If we deduct the discount, $160 from the original price of $400 [400-160] we get the **eBay price of $240**. Therefore, our solutions are correct.

3. Mixture Problems

If you work in a laboratory or hospital, you might well run into this kind of problem solving.

Example 9.

You are a researcher at the Apex Research Facility. You have on hand 4 liters of a solution that is 50% alcohol. Your experiment requires a 40% alcohol solution. In the dispensary, they have a pre-mixed alcohol solution of 20% alcohol. How much of this pre-mixed 20% solution will you need to dilute the 50% solution to make it a 40% solution? How much of the 40% solution will you end up with?

We will start by defining the variables from what is asked in the problem:

"How much of this pre-mixed 20% solution will you need?"
"How much of the 40% solution will you end up with?"

Define the variables: $x =$ **the amount** of 20% alcohol solution
$y =$ **the amount** of the new 40% alcohol solution

First we will **create an equation** that represents **pure alcohol**.

The amount of pure alcohol in the 20% solution can be determined by multiplying the amount of solution by .20 $[.20x]$

We have 4 liters of 50 percent solution on hand. Therefore, the pure alcohol of this solution is $.5 \cdot 4$ liters.

Together this must equal y liters of 40% alcohol. $[.4y]$

This gives us **a basis for an equation based on pure alcohol:**

$$.2x + .5(4) = .4y$$

Our problem is that we only have one equation with two variables. We will have to develop a second equation using our variables, x and y to create a **system of equations** [Topic 14].

The **second equation** will be based on the **total amount of solution** in the problem. Adding x amount of the 20% solution with 4 liters of the 50% solution will equal a total of y liters. This gives us **a basis for our second equation**:

$$x+4=y$$

With two equations in two variables we now have a **system of equations**:

(1) $\quad .2x+.5(4)=.4y$

(2) $\quad\quad 1x+4=y$ [We've added the implied coefficient of 1 to the x variable.]

We can **substitute** equation (2) into equation (1) to eliminate the y variable:

(1) $\quad .2x+.5(4)=.4y$

[Substitution Method – See Topic 14.]

(2) $\quad\quad \overline{1x+4}=y$

This gives us an equation with only one variable which can be solved for x :

(1) $\quad .2x+.5(4)=.4(1x+4)$

To eliminate decimal points in (1) we will multiply the entire equation by 10:

(1) $\quad 10\left[.2x+.5(4)=.4(1x+4)\right] \Rightarrow \quad 2x+5(4)=4(1x+4)$

Distributing results in: $\quad 2x+20=4x+16$

Adding -2x to both sides of the equation: $\quad \dfrac{-2x \quad\quad\quad -2x}{}$

Results in: $\quad 20=2x+16$

Adding -16 to both sides of the equation: $\quad \dfrac{-16 \quad\quad -16}{}$

Results in: $\quad 4=2x$

Dividing both sides by 2: $\quad \dfrac{4}{2}=\dfrac{2x}{2}$

Results in: $\quad 2=x$

Or: $\quad x=2$

Solving for x answers our first question:

We need 2 liters of 20% alcohol solution solution.

Using $x=2$ in our second (2) equation: $\quad 1x+4=y$

Results in: $1(2)+4=y$

Simplifying: $6=y$

Or: $y=6$

Solving for y answers our second question:

We will end up with 6 liters of the new 40% alcohol solution.

Exercises

1. In the formula $y=mx+b$, solve for x in terms of m, y and b .

2. The height of a photo is 2 less than 4 times the width. If the photo has a perimeter of 36 inches, what is the height and width?

3. The number 120 is 60% of what number?

4. Acme TV reduced the price of their $600 model by 20%. What is the new price and how much was the discount?

5. If the $600 sale price of a washing machine is 75% of its original price, what was the original price?

6. If one saline solution contains 60% salt and another saline solution contains 30% salt, how much of each solution is needed to make 33 liters of 50% salt?

Solutions: **1.** $x=\dfrac{y-b}{m}$; **2.** Height: 14 inches; Width: 4 inches; **3.** 200; **4.** New price: $480; Discount $120; **5.** $800;
  6. 22 liters of 60% solution and 11 liters of 30% solution.

This concludes our topics for a basic outline of the principles of algebra . There are many topics that have not been included. Since this book was not meant to be a definitive text, only the most common topics covered in an introductory algebra course have been discussed. Emphasis was put on understanding the basic structure of algebra as well as using this potent extension of arithmetic to analize and solve problems in the real world. This writer hopes that it has been useful for you.

Unfortunately, this writer is error prone. Despite reviewing the book several times to pick up typographical or substantive inaccuracies, there are probably still many here. If you find any, it would be appreciated if you would call them to my attention by emailing me at rmacgregor@smcc.me.edu . Thank you in advance.